COLD FUSION
Dignity of Mind

Marc E. King

Author of *Changing Your Mind* and *Fifth Dimension*

To order additional copies of this book, contact:
Xlibris Corporation
1-888-795-4274
www.Xlibris.com
Orders@Xlibris.com
128079

Table of Contents

An Aside: **Never say "in other words" because then you may be forced to clarify yourself. In that case, your thoughts and ideas would be under review. Forward by the Author**

Foreword by the author

The Introduction to **Cold Fusion** will challenge your thoughts and intuition and will likely lead you to question the higher capabilities of your mind that may be new and very useful discoveries for you.

Previous related works by the author are **Changaing Your Mind** and **Fifth Dimension** as noted in reference. The author has also published a recent modern-health guide related to the scientific advantages shown in the prior works.

While **Cold Fusion** is a non-fiction text, every effort has been made to create a very readable and understandable non-technical view of modern science and technology. This text primarily defines the dignity of mind that is and has been our birthright.

As in the prior works, we cannot be comfortable with rhetoric and common sense alone. We need to technically defend and prove each step on our way. Once we can defend ourselves mathematically and physically, then much of the criticism becomes superficial, and the obstacles in our path remove themselves.

Most technical material has been confined to short appendices at the end of the main text.

Some diagrams and some arithmetic have been offered in the main text as a conceptual tool to be helpful or to be ignored as the reader desires.

While the appendix technical material accomplishes more proofs from, and more additions to, the original technical establishments of the prior works, it is by no means required for the reader to appreciate the simple language version of this book.

Technical readers who desire to follow in sequence can simply turn to the appendices, but it is recommended to read all the material.

"Cold Fusion" is a model for the technical concept having the same name, but it is more importantly an extension of that concept to further use the same low energy science for communication purposes.

Introduction

Imagine yourself most recently in the mirror.

Imagine the entrance to the place where you live or work. Next, imagine a close friend or relative. Next, imagine a summer scene and then a winter scene near where you live. Now imagine your favorite sports team, favorite musician(s) and your favorite necktie or outfit. Also imagine the place where you buy groceries and also its parking lot or street.

You have most likely just now reviewed each remembrance as a photo-type image in your mind. Most were in black and white and a few were in color.

Imagine you were once in a classroom where you were not interested in the subject matter.

Did you have other thoughts and imaginations that did not relate to the classroom or subject matter at hand? Did they also seem photo-like in part?

Have you ever had a vivid dream where the visual resolution during the dream was quite descriptive? Have you ever heard a voice in a dream?

Have you ever had an experience while awakening from a deep sleep where you seemed to be coherent and able to rise but you were not able, at that moment, to move your body or even speak (kind of like your cell-phone or laptop "locking up" so that you needed to power-down-restart and re-initiate everything?)

Or experienced a dream where you were able to "participate" in the dream, for example interacting with characters during the dream?

Have you ever had a premonition that something just "wasn't right?"

Have you ever felt as though "I have been in this scene before!"

Have you ever met someone that you felt extremely comfortable with or that you immediately felt uncomfortable with? Have you just now reviewed a photo-like image of such an experience(s)?

Have you ever wished that you were somewhere else and viewed the place you would rather be as a sort of photograph?

Have you ever thought about your workplace on the last day of a weekend and viewed, in your mind, a recent scene at your workplace or an image of your boss?

If you think about the nearest window to you right now, what do you imagine to be viewed outside (or inside) that window? Do you imagine that view as a sort of photographic image?

The point of the questions is that you have likely answered yes to many or all of them.

Your mind is your birthright and has fabulous capabilities for your exploration.

"The common perception of the brain is that it is something like a huge magic super parallel processor. That it comes equipped with infinite ROM (read-only-memory) and even a little communication ability as part of the bundle.

Nothing is farther from the truth.

To start, please think about any important memory, e.g. a high school prom, your first remembrance, a first date, a favorite teacher or professor, yesterday morning, a friend or relative, anything at all.

A photographic type image immediately appears to you for the memory you have addressed. There is no end to how many of these photo-type images you can visualize. Basically, an infinite number of images.

Do you truly believe there is a magical infinite ROM (read-only) memory storage within the size of a grapefruit sitting above your shoulders? Infinite? Never ending? The size of a grapefruit?

Most likely not."

III. Changing Your Mind

In the text **Changing Your Mind**, it is suggested that time as we know it is an excellent measurement tool for timing our days, nights, seasons, years and much more.

It has now been proved by several means that the concept of continuous time is a wonderful perception and measurement device (clocks) for our daily use.

We also now understand that the perception of time, however useful, breaks down under scrutiny.

Time is in fact different from the way we perceive it to be. It is not real, and it is not mathematically continuous. It is more like the still frames that make up the "motion" of a motion-picture than the continuous view of the motion-picture itself.

Time is a wonderful concept but is truly only an approximation and representation of macroscopic reality. Macroscopic reality is important! But while it defines our daily lives, it does not define the exact principles needed for our electronic devices (cell phones) and everything else we use that is based in electronic "quantum-science" (well known since the year 1920.)

Current science (since about 1920) shows that time is somewhat mysterious regarding very small (less than microscopic) and very large (beyond our solar system) objects. The prior work *Changing Your Mind* dispels the "mystery" and presents understandable new science backed by mathematics and modern physics.

After reading *Changing Your Mind*, the reader now understands that the "old" version of modern physics was greatly useful, but simply does not apply in the year 2012 and beyond.

The non-technical version of *Changing Your Mind* has been attached as a non-technical reference for the concepts of the spatial sequential progression. Many concepts are geometrical in nature, and many readers have found the intuitive geometry shown in the non-technical text to be very useful!

An Aside:

Anyone in the US who has traveled frequently to and from Asia has a head-start on the fictitious nature of time.

While we depart, typically from the West Coast and arrive in the evening of the next day (crossing the International Date Line,) similarly we depart Asia and then "arrive" back on the US West Coast effectively "before we left."

The best (most unfortunate) part for those who work in Silicon Valley is that we have then arrived in time to get to our workplace in order to contribute twice for the same day!

Everyone understands the International Date Line, but this example shows the somewhat fictitious nature of our so-called real-and-continuous time t.

PROGRESSION OF THE SPATIAL SEQUENCE (APPENDIX P FROM FIFTH DIMENSION)

The concept of continuous "time" along with the reality of spatial progression is represented in Diagram 1; <u>One Dimensional Spatial Progression Model</u>.

While the sequence moves directionally in three dimensions through five dimensions (the blue arrows,) we perceive only the 3-dimensional progression (green arrows) due to our limited synaptic "speed" or rate-of-perception.

The "speed of light" (denoted c) is known to be 300,000,000 meters per "second" of our perceived time t. The speed of light is much greater than the fastest space ship or satellite velocity and represents the distance b (one second of perceived time divided by 300,000,000 or a tiny fraction of one second.) The metric b is shown in the diagram of sequential spatial progression.

The distance b can be thought of as the smallest or most precise possible movement through space. The distance b can equally be thought of as the smallest unit of measure on a ruler or other measuring stick, e.g. as one millimeter is a small segment of one centimeter and a very small segment on a meter-long measurement stick, and as one mil (thousandths of an inch) cannot even be discerned on a yard-stick segmented by increments of feet, inches, quarter inches, eighth inches, and sixteenth inches.

While the extreme value of the speed of light c allows for a very small increment along a one dimensional "meter measurement stick" (the value b,) our vision and our entire neurological and synaptic ability allow only for the minimum perceivable increment of about 1000 x b. For example, we can see the inch line on a ruler but we cannot see the one-mil "line."

In other words, we do not have and never will have the ability to perceive the higher dimensions even though we exist arguably 50% in dimension D=5 and above and the other 50% in our perceivable dimension D=3.

While our minds understand the fifth dimension (memories, analytical thought, dreams and more,) our sensory perceptions do not allow a clear and visualized understanding of higher dimensions.

IV. The Fifth Dimension Made Easy

The concept of five dimensions is not mysterious.

We live in a three-dimensional world where we can travel 1) right or left, 2) forward or backward, and/or 3) up or down. 3 dimensions.

If you have ever been on an airplane, then you already understand the idea.

Living in three dimensions, we already understand two dimensions (right/left and up/down.) This is the way we view a blackboard, a chalkboard, a white board with colored markers, a TV screen, or our cell phone two dimensional display.

Even when we try to imagine three dimensions on a two dimensional page, we use imaging enhancements such as shadowing, etc. to emulate a three-dimensional view.

An example is to view Diagram 2.

The concept of five dimensions is simply the following:

$3 + 2 = 5$ where we already understand 3 and 2.

Dimensions higher than five, for example 8?

$8 = 5 + 3 = 3 + 2 + 3$ where we already understand each and every constituent.

There is no true mystery.

While it sounds "weird," it is in fact "straight forward."

Diagram 7 shows a figurative model of 8 dimensions that can be visualized by what we already understand, i.e. $3 + 2 + 3 = 8$.

The *Fifth Dimension* text, among other things, points out the clear nature of artificial logic (known as electronics as in your personal cell phone.) Instead of simple electronics (meaning the electron nature of atoms and molecules,) the non-electron lower energy parts (nucleus/nuclear charges) of atoms are also included in new fifth dimensional physics.

These charge combinations effectively add up to $3 + 2 = 5$ while we have so-far been using only $3 + 0 = 3$.

This is not a great leap into misunderstood (weird) dimensional space; instead, it is the same addition-arithmetic we have all learned in grade school.

The concept is very much like an earth worm moving in a single dimension as the worm comprehends. As the worm cannot see, but we can, the worm is not moving in its own one-dimensional straight-line progression. It is in fact moving right/left, forward/back and even up/down (three-dimensions) underneath the earth (surface dirt.)

The worm "thinks" (non-arithmetically!) that $1 + 0 = 1$ while we can see that the worm is in fact experiencing $1 + 2 = 3$.

While experiencing 3 in the absence of understanding or acknowledgment for $3 + 2 = 5$, then we are similarly existing in dimensional space D=3 with no understanding of higher dimensional reality.

In a similar and previous manner, without acknowledging even 3 dimensional space, we have behaved like two-dimensional creatures believing the Earth surface was flat as opposed to in fact curving in the next higher dimension 3.

The five-dimensional-computer model (electron + proton) has been shown and understood. The machine can operate extremely faster than the standard existing (old) computer or cell phone model.

Technically, carrying forward from the Mass-Space results from *Changing Your Mind*, *Fifth Dimension* defines the Charge-Space requirements that define the periodic chart of natural element, the five dimensional computer, and also the peaceful cold-fusion model for low cost energy.

Appropriate excerpts from *Fifth Dimension* are attached for reference in the appendices.

V. Dignity of the Mind

BASEBALL AND SOFTBALL

From a US perspective, baseball and softball players and coaches know that the batter needs to keep his or her eyes on the ball.

While the pitcher throws the ball so that it appears to be approaching at a given speed and with a given direction, the ball in fact moves faster or slower than it appears to the batter and it may easily move up, down, right, or left as it nears and crosses the plate.

A good pitcher can fool many batters while a good hitter can take advantage of many pitchers.

This is the nature of the game(s).

Throwing a pitch and hitting a pitch is best enjoyed by watching the game; however, these athletic actions are closer to the true facts of physical science than you may have previously imagined.

Similarly, from a worldwide perspective, a football (soccer ball) kicked or headed toward the goal may be approaching fast, slow, right, left, up, down, and/or may be spinning in different directions. The goaltender needs to quickly figure out the spatial approach and approaching manner of the ball.

In higher level sports competition, movement in fact exceeds the perceptual ability of the responder, but the responder is able to respond anyway.

MODERN THEORY
OF OUR COMMUNICATION ABILITY

As shown, the concept of time is an approximation (a very good approximation) of day-to-day experience; however, the concept breaks down under mathematical and physical scrutiny.

In particular, time "t" is *not* a real and continuous entity (variable) in scientific and mathematical terms.

This fact is proved in the manuscript *A Mathematical Transformation of Variables* and the manuscript has been attached in the appendices for reference.

In the modern world, the concept of continuous time is replaced by contiguous energy-loss through sequential spatial traversal.

Among other things, this leads directly to the concept of (quantum) energy loss for our hydro-carbon molecules of life and creates an "aging" condition that seems "unstoppable" for hydro-carbon life.

Modern science now understands that "continuous aging" is not true.

SUMMARY OF RESULTS
FOR HYDROCARBON MOLECULAR LIFE

In the modern three-dimensional world, molecular "time" is obsolete and our living hydrocarbon molecules decay at a quantum energy rate ($680eV$ kg^{-1}) per sequential boundary assuming the exact quantum sum is available and achievable.

In case the exact quantum sum is not made available, then the hydrocarbon molecules of life cannot decay based on quantum physics that has been understood and acknowledged since the approximate year 1920.

The methods to prevent the exact quantum sum are prolific[1,2].

THE BRAIN'S COMMUNICATION ABILITY

We all grew from (and consist of) the natural elements of Hydrogen and Carbon as in our hydro-carbon molecules of life.

Hydrogen is our most easily available means for communication through low energy electro-magnetic "waves." These waves have been called, in a very different context, "brain waves."

Diagram 3 "Forward and Reverse" is a figurative and physical one-dimensional attempt to show low energy (infrared/low-thermal) waves propagating in "both" directions (forward and reverse.) This is the communication process for the forward direction and for the "previous or backward" direction as below. (If Diagram 3 is useful, then internalize. If it is not useful, then ignore.)

Within the brain, Carbon is a building block upon which to mount and connect geometrically precise Hydrogen atomic positions within life-functional (operational) hydrocarbon chains of molecules.

The small energies ("fine" and "hyper-fine" spin-split energy states) of Hydrogen are asserted means for universal communication per Diagram 12; Spin State Split Example for those who desire to review. This diagram is technical and is not required except for symmetrical (geometrical) intuition.

The energy state transitions are so small (micro-eV) having wavelengths of several to tens of cm, that the deep infrared (IR) thermal energies are continually excited in the same way that your body feels warm when in sunlight, when near a campfire, or in the warm indoors on a cold evening.

For all life, the excitation of living hydrocarbon molecules to "communication wave" stage is then continuous and the low energy waves "always" propagate.

(Note: The diagram "Forward and Reverse" is a right-left (one dimensional) view of a three-dimensional world (a fixed atom in its own origin of coordinates for now) and is difficult to visualize, but the concept does not change for three-dimensions.)

The brain then has a continuous five-dimensional ability to communicate along the direction of the forward spatial sequence and/or along the reverse spatial sequence. The results:

1. *FORWARD:*

Analysis, other processing thought, dreams, and much more is processed, per the diagram, effectively "in-between" our three-dimensional intervals of reality. The brain has far more dignity than we have ever imagined.

Think of turning on the room light switch. Between the "time" you hear the switch click "on," and the "time" the room lights illuminate, your brain has processed about 300,000,000 thoughts.

The higher-dimensional Hydrogen nucleus charge motions are not independent; instead, they inter-react with each other in order to "think" at a very fast rate (c^2) while the rest of the brain assimilates and operates at a more standard rate (c) and in fact at a rate smaller than c ($\leq c$) due to the slower operation of the chemical nature of our biology. (The rate c represents the "speed of light.") Diagram 18 shows the relatively large number of combinations provided in five-dimensional space.

A crude example of the higher dimensional (nuclear or "magnetic") interaction is shown in the low-energy-computer "Bit Map" diagram from Appendix S (Diagram 17) for those interested.

2. REVERSE:

ROM (Read-Only-Memory) is a result of the brain's ability, again as viewed in the "Forward and Reverse" diagram and also Diagram 6, to address prior (previous) spatial frames that are in a "backward" sequential position.

Once your brain has been instructed (by you) to address a prior or past frame, for example the image you immediately "see" when you think of any previous school or classroom, the communication to and from the prior frame takes place at the fast rate (sequential rate c^2) and then back through your (slow) chemical synapse for you to "view" as quickly as possible once you have addressed it.

From reading the above paragraph, you may have envisioned a classroom, a teacher, or a setting from one or more schools from your past recollection.

The information in the frame(s) you just now viewed was not somehow captive and stored inside your head as you may have previously thought; instead, your brain has the dignity of traversing to the prior spatial frame and returning the information to you at your command.

While the "five dimensional computer" model in Appendix S has severe limitations in "reading" (distinguishing) the small energies involved using transitions from five-to-three-dimensional space, your mind has the dignity of unlimited ability regarding higher dimensions.

A child's magical ability to transition into language and then to multiple languages is not in fact "magic." It is the inherent, automatic and autonomic ability of the brain's thermal (low energy wave) communication process (mechanism) as it grows to become fully mature (operational.)

This is our birthright.

Building upon simple new building blocks, we have now achieved a higher level from which to see.

An Aside:

The author's first day in Kindergarten was disconcerting.

On the bus to the very first day of school ever, the author looked out the window and watched the city go by. He thought of his neighborhood friends and of walking on the front walls together every day and having fun.

He thought of where each of his friends was going to school on that same day. The author thought that life as he knew it was over. Completed. Further destined to an eternity of daily ritual. Forever. No reprieve.

The author was right.

Yet it seemed the circumstances should somehow have a higher meaning. A more important cause than to educate, work, contribute, proliferate, and then die.

Even through his religious education, it made no sense that God would create all of the physical universe unless the exact physical universe was necessary. Otherwise, everything was some kind of grand joke or mistake?

Grand joke or mistake has been rejected in this model.

A Short Technical Aside:

Interface from Five to Three Dimensional Space

Using the crude example of five-dimension-computer principles to suggest ways to help understand the brain's ability and method, we immediately see the following from the results of Appendices Q and S:

1. *The input or "write" energy required to the machine is only a function of the desired proliferation of the low energy progression "through" the machine. This should be straight-forward.*

2. *Any output (back to three-dimensional energies) should be quite difficult.*

Some of the existing ways our three-dimensional science has dealt with low energy enhancement for the purpose of logical value improvement and recognition are shown in diagrams of common power-amplifiers not dissimilar to the amplifiers for existing ground-to-ground communication signals.

THE COLD FUSION MODEL (OF LOW ENERGIES)

THE UNFORTUNATE HISTORY
OF NUCLEAR FORCES

In the worldwide conflicts of the previous century, Mankind revealed its most brutal face.

The face of aggression and reply lead to the science of nuclear weapons.

Although a simplistic scientific means per our present time, the large forces of the atomic nucleus were released upon the world as a weapon and have been known as a weapon ever since.

The nuclear forces were discovered and used in a primitive means of basic statistical understanding for unleashing the strong forces that maintain positive nuclear charge in a close proximity, i.e. the small atomic nucleus.

We now understand that the atomic nucleus is in fact five-dimensional in terms of energy capability, while the atom itself has different quantum electron energy positions and is three-dimensional.

We also know that three-dimensional "time" itself is in fact a representation of the quantum energy requirement through our spatial progression. For example, our own hydrocarbon molecular structure does not decay as a function of time; instead, our living molecules and cells decay through the quantum energy requirements defined by our radial (from center-of-mass) position on the planet surface as a function of mass and space (gravity.)

While the atomic nucleus has remained a mysterious and powerful misunderstood reality, it is also a source of life itself and is capable of providing energy and thought in a peaceful and extremely productive way.

RADIOACTIVE (HOT) FUSION

In 20th Century science, in an expedited manner to create weapons, the radioactive means for "splitting" unstable atomic nuclei ("fission" weapons) was developed in order to unleash power contained in the form of high energy and strong forces from the atomic nucleus.

This was effective as a powerful weapon but was quite crude and unproductive in a relative way.

Soon, the concept of "fusion" weapons was developed along a similar principle, i.e. instead of "splitting" an atomic nucleus, the attempt was made to perhaps "fuse" an atomic nucleus into place with the existing strong energy known as a "fission" reaction.

The attempt worked and it was shown that the radioactive "fission" process could supply enough large random statistical energy to induce a "fusion" reaction that was extremely exothermic (generating extreme heat as in a thermo-nuclear weapon.)

With the new primal empirical knowledge, scientists began to develop a productive theory of energy production using the extreme power of the atomic nucleus.

Unfortunately, the fact remained (at the time) that "fusion" (clean and harmless if used correctly) could not take place without the relatively "dirty" (radioactive) process of "fission" as in the very old style weapons.

Nuclear power plants today, all over the world, have the extreme danger resulting from lack-of-control over a statistical and dangerous semi-controlled reaction.

COLD FUSION

Cold fusion is a means of generating an energy producing nuclear reaction that does not require the dirty and radioactive process of "fission weapons."

While the fission reaction provides enough energy to "fuse" a nucleus together using crude statistical means, the "fusion" reaction itself is much simpler and more precise (controlled) than the old "fission-weapon" technology.

Using the exact same principles, in a different way, as the five-dimensional-computer (nucleus charge enhanced processing speed,) the cold fusion reactor can supply effectively infinite energy for 100% of our collective needs without danger of crude radioactivity.

Most of us have held two opposing (same pole) magnets in our hands during grade school science class and have felt the repulsion (repulsive force) preventing us from combining the magnets in space.

In a very similar way, the positive (nuclear) charges provide a subtle force of interaction in our daily world.

For very low cost, those same nuclear forces are fully capable of providing more ***controlled*** energy than our existing three-dimensional world has ever imagined.

In that case, dead dinosaurs should truly be obsolete and completely worthless.

An Aside:

The universal fact of low energy (transmission) waves is required for the brain to communicate. The energies are not present by "mistake" or frivolity.

Our communication backward (memories) and forward (analytical and creative thought) are our sanctity of mind. This communication is our dignity and our birthright. It represents a responsibility.

VI. Another View of Low Energies

As we have discussed, the nuclear (positive charge) energies are very small relative to our three-dimensional experience.

But we need to be clear.

The energies are "small" (being distinguished by close increments) in scientific terms.

That does not mean that the summation or total value of nuclear energy is small. In fact, the opposite is true.

The forces (related to energies) holding the atomic nucleus "together" are extremely large by our three-dimensional experience.

The nuclear energies, being small in value relative to our three-dimensional experience, sum to large values of small energies, i.e. low-energy "thermal" large energies similar to the relatively low energy of a campfire or barbecue that has become a multitude of fires that have consumed several mountains and other expansive terrain.

Small energies are capable of large effects.

As with technical "cold fusion," we would prefer having a living room fire to warm us thermally rather than having a large "quick" release of massive thermal energy that would serve only to disintegrate all matter within its range.

The productive and controlled method of cold, as opposed to hot, fusion then becomes a valuable asset as opposed to the previous counterproductive "asset."

The technical appendices review these methods for those having interest.

VII. *In Between the Light Switch*

The faster concept (faster than the speed of light) is discussed in sections VIII and IX.

The dimensional facts assert the speed of light (c) is quite "fast" in our three-dimensional perception.

We have already shown (technically) that the speed of light c can be greatly out-paced.

From the time you turn the light switch "on" and the time the room lights illuminate is a very long time in dimensions higher than three.

This is in fact the principle of the five-dimensional computer (5D PC.)

In this case, "under the radar" takes on a new meaning.

VIII. Getting Ahead of Yourself

The fifth dimension only sounds mysterious. It is not.

There is no mysterious spot in outer space beyond some "black-hole" labeled "fifth dimensional space."

Fifth dimensional space means an environment where energy requirements are very low relative to what we ordinarily experience.

For example, imagine you are in the perfect-space-suit and you are floating in between the Earth and the Moon. You would be influenced by very little gravitational force. You would spend basically zero energy to move your hands up and down because there would be nothing in fact "pulling your hands in either direction."

This concept is very close to the higher dimensional concept that requires little energy. No one needs to be between the Moon and Earth or anywhere far away from mass (as in deep empty space) to experience low energies. (But you would experience very low energy requirements if you were there.)

LOW ENERGY SPACE

The atomic nucleus (part of every atom that makes up our living constitution) is known to have a high spatial-bounding force, but it has very low internal energy requirements.

The atomic nucleus is "five-dimensional" in terms of its required internal energy.

As one example, the "five dimensional computer" does not exist in outer space or in a secret closet.

It exists along side of us just as a silicon chip exists in our cell phones. The difference is that the "faster" computer utilizes the positive (low energy nuclear) charges for operations (computations or communications) in addition to the standard higher energy-requirement electronic charges that we have been acquainted with for decades and centuries.

The 5D PC is simply technology moving forward and is no more "weird" than first using a light bulb instead of a candle or simply taking a photograph instead of making a drawing.

Because the five-dimensional computer uses proton (positive) charges, it has also been referred to (inaccurately) as the "magnetic" computer.

A conceptual way to view the 5D PC structure is simply to acknowledge the extremely low energy requirements for the positive charges to "move." Effectively, while electron (negative) charges are "moving" under large three-dimensional energy requirements, the proton (positive) charges "move" for *free* relative to three-dimensional (standard/old) energy and cost requirements.

At each "electronic bit" change, the positive charges have already been able to process logical operations "in-between" the slow electron movement.

The five dimensional computer is presently the best emulation (in a crude and humble way) of the capability and method of hydrocarbon brain function thought, analysis, and imagination.

The two added vertices (degrees of freedom) in "five-dimensional" space allow the convenience of thinking, analyzing, creating (processing) "ahead of" the limited three-dimensional capability we already posses.

A Real Life Aside:

Have you ever been in a warm and comfortable place (perhaps in bed dreaming) and then had an uncomfortable sensation, for example that you were falling? What about falling or diving from a steep cliff while standing on the edge and looking down 200 feet? What if you knew the water below was very cold?

Have you then experienced a sense of "height" or a sensation of falling, or even felt the cold water upon your otherwise warm skin?

You are able to feel these feelings in an "instant" because of the extreme imagination and thinking (processing) speed of your mind.

IX. *Arriving Before You Get There?*

Suppose we left for a dinner party and when we arrived we were already there and had finished dinner!

If we had access (we do not in this context) to the extra 2 vertices per above, then we could arrive at any party "ahead of ourselves."

The reason is that we would not need to take the "curved" route that everyone (including us) needed to take; instead, we could simply take the "straight-line" route as shown in the example figure in Diagram 6.

While this is a figurative shortcut, in fact our three dimensional perceptions view the curved line as being straight.

We could only see and experience that the 3-dimensional "straight-line" is in fact crooked after we have understood the fifth-dimensional "straightness." Then, we would naturally take the truly "straight" path.

A Figurative Aside:

Pretend your job is to be a stock trader on Wall Street. You and your neighbor, best friend, and fierce competitor both leave your building at the same time to get to the office for the start of the trading day.

You each take different buses. Your friend's bus is non-stop to your workplace. You take a different bus that has many stops.

Cell phones do not make any difference on these buses. A bus rider can only achieve physical information at the stops.

The buses arrive together at your workplace (identical time-of-morning) even though your bus made many stops. The reason is that your bus driver took a route of dimensional travel and knew how and where to get information (using dimensional rules) and your neighbor's bus driver had taken a more conventional route.

Once you and your neighbor arrive at work together, you already posses all knowledge from events during the journey while your fierce competitor (best friend) has only the same knowledge as when he or she left home.

In this case, you will likely have a better day at work than your neighbor.

In each next dimension, we can see a straighter path between points.

For each dimension, the shortest distance through space is reduced.

Example 1:

If we lived in one-dimension, we would not be aware of the curvature of our single dimension (a line) through the next higher dimension (a two dimensional plane as the surface of the page.) While we may travel a route that seems completely straight to us in one-dimension (the curved and longer distance following the arc in Diagram 6,) someone with knowledge of two-dimensional space would prefer to take the "straighter" straight route (the straight-line distance.)

Example 2:

The Diagrams 15 and 16 show the curvature of the planet Earth as a two-dimensional surface that is "curved around back onto itself" through the next higher dimension (the well known third dimension.)

If we lived in the year 1492, we would probably think the Earth surface was flat. If we took a sailing ship from London or Plymouth to Florida or from San Francisco to Hong Kong using the "flat" map, we would be traveling a greater distance than the modern day airplane pilot who takes the geodesic route that follows the curvature of the Earth surface and achieves a "straighter" straight line from start to finish.

While it's not as easy to visualize, the five-dimensional route from A to B is much shorter than the three-dimensional route in exactly the same way.

We all have a feeling for the word infinite (never ending.)

As the higher number of dimensions becomes much larger than the numbers 1, 2, or 3, there would be no limit to the "straightness" we could achieve for the geometry between points A and B.

In other words, there would be no limit to the number of intermediate no-time-required "bus stops" and there would be no limit to the shortness of travel from A to B.

In the Wall Street bus stop model, three-dimensional space is achieved through five-dimensional space "sooner" than the three-dimensional arrival.

In the infinite model, three-dimensional space should be in fact "already achieved."

Examples A and B:

The following figures (Diagram 4) show two ways to visualize "already-achieved."

Example A

Imagine you are at the center of a one-dimensional (closed-*line*) circle. Then you are equally distant from each point on the circumference (perimeter) of the circle and you are as close to any part of the "circle 1-D universe" as you are close to any of its other universal points. Nothing in that one-dimensional space is closer or farther away from you than anything else.

Now, imagine per the diagram, that the circle gets closer and closer to you (its radius R is decreasing.)

In the limit of very small R, the circle would be indescribably close to you since you are at the center.

In that case, you would not only be equally-distant from all the linear (circular) points, you would also be indescribably close to all of them.

You would effectively be everywhere in the one-dimensional universe at once. (Everywhere at the same "time" as far as the circle could experience.)

Example B

A more realistic example:

Now imagine that you are a higher dimensional creature (two-dimensional instead of one-dimensional in this example.)

You are kind of like a ground-creature moving along a closed surface (the ground surface of the planet Earth.) You think the Earth is a flat surface.

You cannot fly as a bird and have no idea there is in fact a third dimension that goes up-down as well as forward-back and right-left.

While you can easily see one-dimensional creatures from your perspective (earth worms, etc.) you spend most of your time dealing with other two-dimensional creatures (that may want to make you part of their food chain.)

Dimensionally, a closed lower-dimensional geometry (e.g. a one-dimensional circle) is trivial for your two-dimensional capabilities. You can easily expand your actions (space) to include the entire area within the closed circle. (Perhaps make a one-dimensional creature part of the food chain instead of you!)

Similar to Example A, once the surface area (two-dimensional) meets the one-dimensional circle, then the two-dimensional experience is effectively "everywhere" within the one-dimensional experience at once (at the same "time" from the one-dimensional perspective.)

If we imagine the one-and-two dimensional circles from Diagram 4 to be two-and-three dimensional spheres, then we see a similar view between two-and-three dimensional space as in examples A and B for the lower dimensions.

A model of our 3-dimensional universe in the next higher (natural) Fibonacci[3] dimension (D=5) is shown in Diagram 5.

As in the above examples A and B, imagine the radius **Rxy** in Diagram 5 to become very small as in Example A.

In that case, the three-dimensional spatial "circular" sequence as shown would be "compressed" so that the spherical (3D) figures became closer and closer together until they effectively merged. The circular "start" would not only "end" at the other side of the "start," but all points (positions) in the three-dimensional universe would be extremely close to each other as in the above one and two dimensional examples.

From a higher dimensional perspective, three dimensional space would be "closed" (already arrived at the "back of itself" through five dimensions.) Its curvature would be 1 as in the one-dimensional line that "closes" into a circle instead of moving in a new direction. (The line has "run into the back of itself" to form a circle.)

Instead of the three-dimensional figures moving in a straight and never ending line (curvature 0) or the three-dimensional figures moving along the natural growth sequence in a spiral (having a curvature in-between 0 and 1,) the closed form figures (curvature 1) do not go to any new place. They simply return to where they started.

In the fifth dimension, the start-to-finish of the three dimensional sequential universe should already be achieved.

X. Conclusions

As a flat Earth surface becomes a round planet Earth, we can now understand the newer dimensions.

Building upon modern building blocks, we have constructed a higher level from which to see.

What our minds inherently know, our perceptions can now learn.

Our mind is our birthright.

Its dignity is universal.

While we access prior spatial frames to remember, we think in forward spatial frames to create.

TECHNICAL CONSIDERATIONS

While the additional technical aspects of low energies, Cold Fusion, are provided in the following sections, they are not required for the purpose of this text.

From modern science, we now understand the true nature of time and the nature of the related spatial progression.

From previous works, we understand the nature of aging (quantum energy loss) and the nature of the brain's communication ability through a lower-energy space than we can imagine in our everyday experience.

To use new terminology, our 3-dimensional biological and chemical processes are slower than our 5-dimensional thought, analysis, creativity, and memory.

We now better understand the nature of ourselves.

Our Hydrocarbon molecules of life are made of atoms as is all matter (substance.) We now understand that the chemical (electronic) nature of atoms requires a relatively high energy in order for interaction (to move or operate) while the atomic nucleus requires far lower energy and acts and/or responds faster than our relatively slow three-dimensional chemistry.

This is the nature of our mental ability.

This is the reason we can think far ahead of when we can three-dimensionally write-down a thought or even speak it. This is our birthright and is our unique dignity of mind.

XI. Diagrams

THIS TABLE IS A REFERENCE FOR ALL DIAGRAMS:

Diagram 1. The Spatial Progression

Diagram 2. Figurative Model of 3-Dimensions + 2-Dimensions

Diagram 3. Forward and Reverse (Photons)

Diagram 4. Examples for 1 and 2 Dimensions

Diagram 5. Example for 3 and 5 Dimensional Closure

Diagram 6. Figurative Model for 3 within 5 Dimensional Motion

Diagram 7. Figurative Tangent Model for 3 + 2 = 5

Diagram 8. 5D Computer Schematic Model

DIAGRAMS

Diagram 1. The Spatial Progression

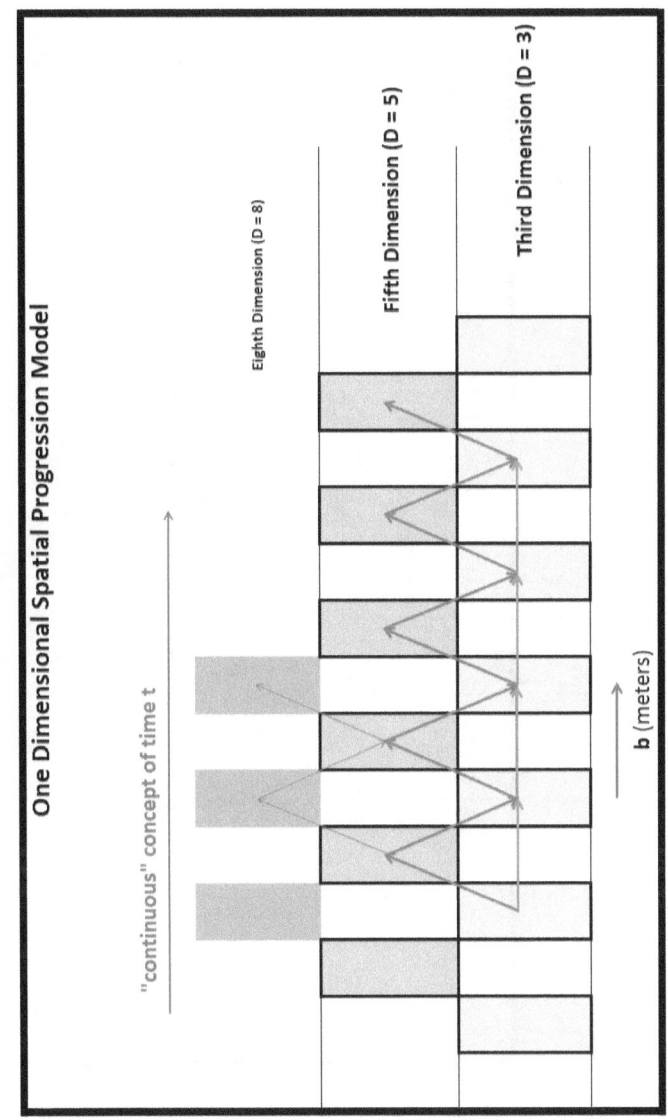

One Dimensional Spatial Progression Model

"continuous" concept of time t

Eighth Dimension (D = 8)

Fifth Dimension (D = 5)

Third Dimension (D = 3)

b (meters)

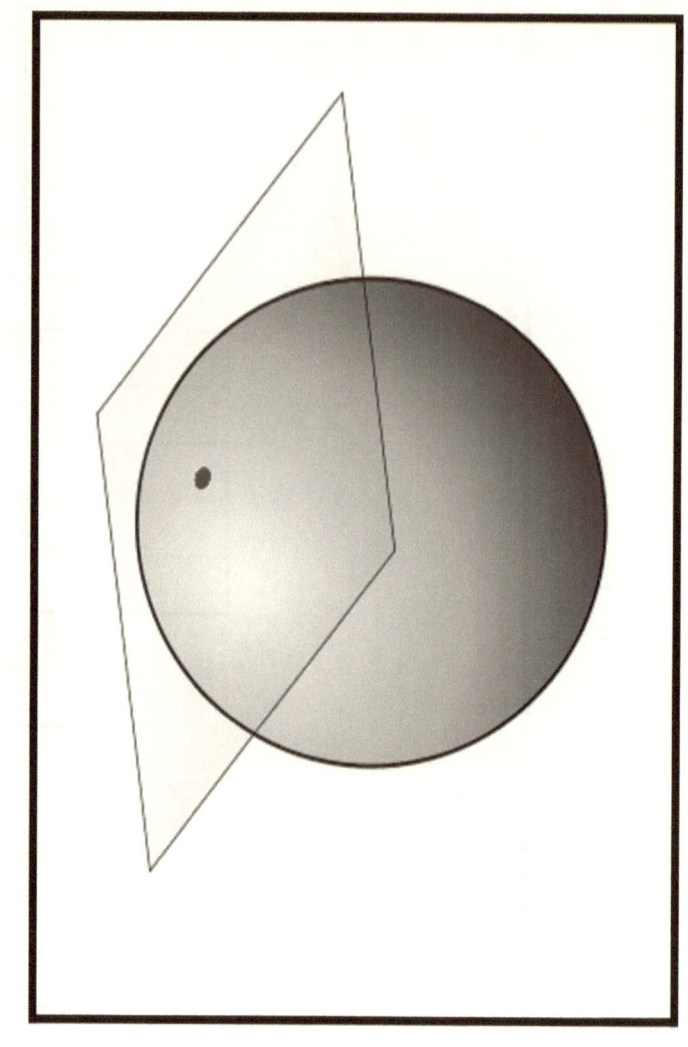

Diagram 2. Figurative Model of 3-Dimensions + 2-Dimensions

Diagram 3. Forward and Reverse (Photons)

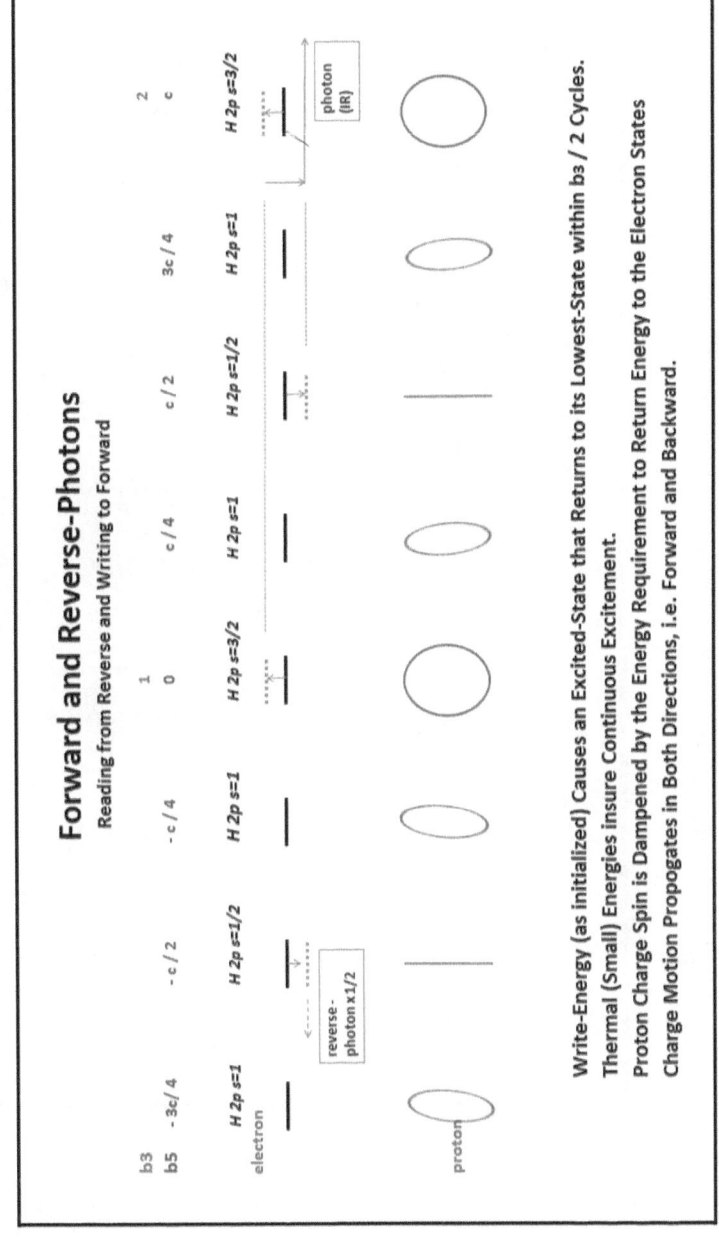

Forward and Reverse-Photons
Reading from Reverse and Writing to Forward

b3	- 3c / 4	- c / 2	- c / 4	0	c / 4	c / 2	3c / 4	
b5			1					2
								c

| H 2p s=1 | H 2p s=1/2 | H 2p s=1 | H 2p s=3/2 | H 2p s=1 | H 2p s=1/2 | H 2p s=1 | H 2p s=3/2 |

electron

reverse-photon x1/2

photon (IR)

proton

Write-Energy (as initialized) Causes an Excited-State that Returns to its Lowest-State within b3 / 2 Cycles.
Thermal (Small) Energies insure Continuous Excitement.
Proton Charge Spin is Dampened by the Energy Requirement to Return Energy to the Electron States
Charge Motion Propogates in Both Directions, i.e. Forward and Backward.

Diagram 4. Examples for 1 and 2 Dimensions

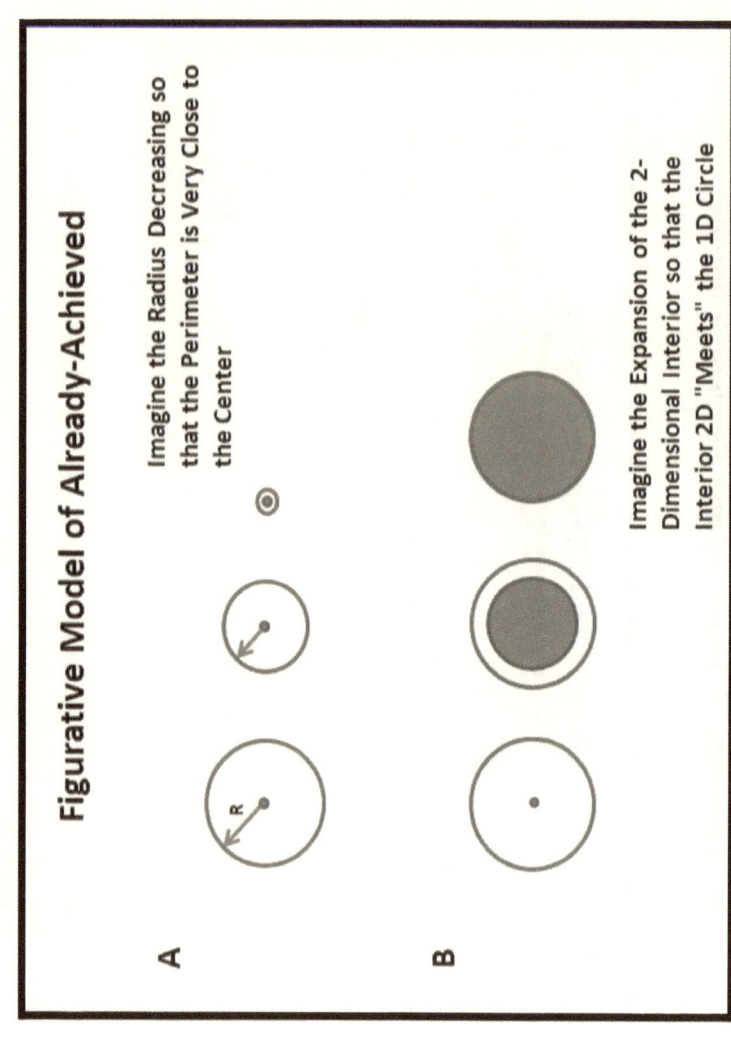

Figurative Model of Already-Achieved

A

Imagine the Radius Decreasing so that the Perimeter is Very Close to the Center

R

B

Imagine the Expansion of the 2-Dimensional Interior so that the Interior 2D "Meets" the 1D Circle

Diagram 5. Example for 3 and 5 Dimensional Closure

Figurative Example of a Closed 3D Universe

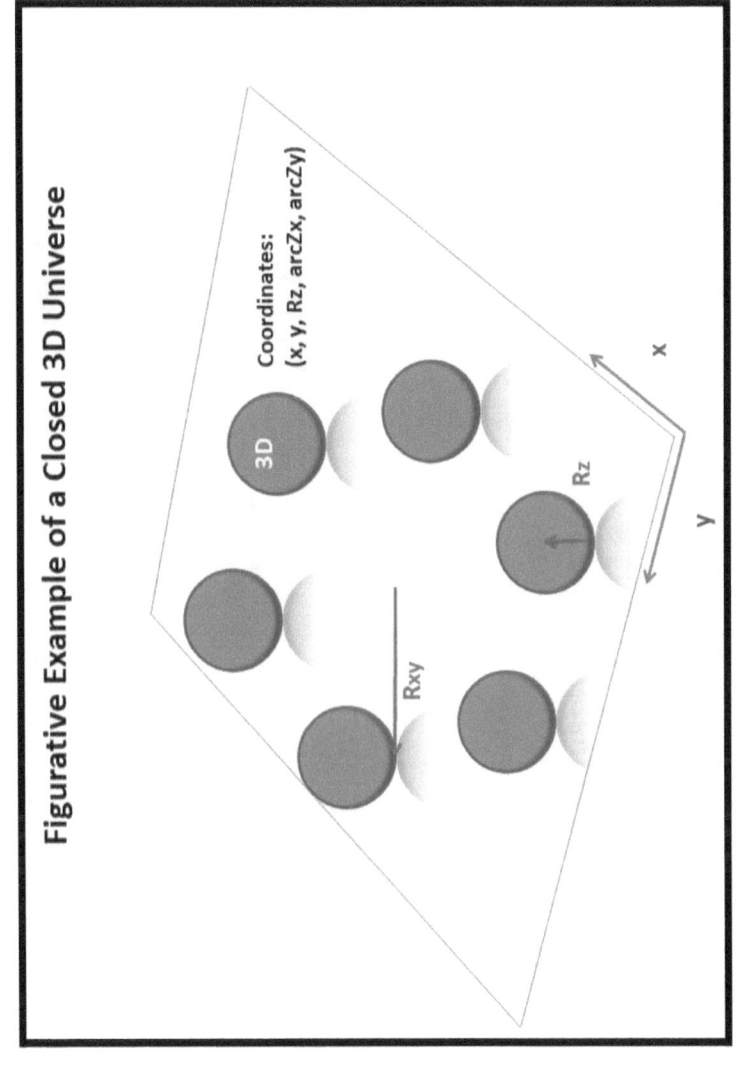

Coordinates:
(x, y, Rz, arcZx, arcZy)

3D

Rz

Rxy

x

y

Diagram 6. Figurative Model for 3 within 5 Dimensional Motion

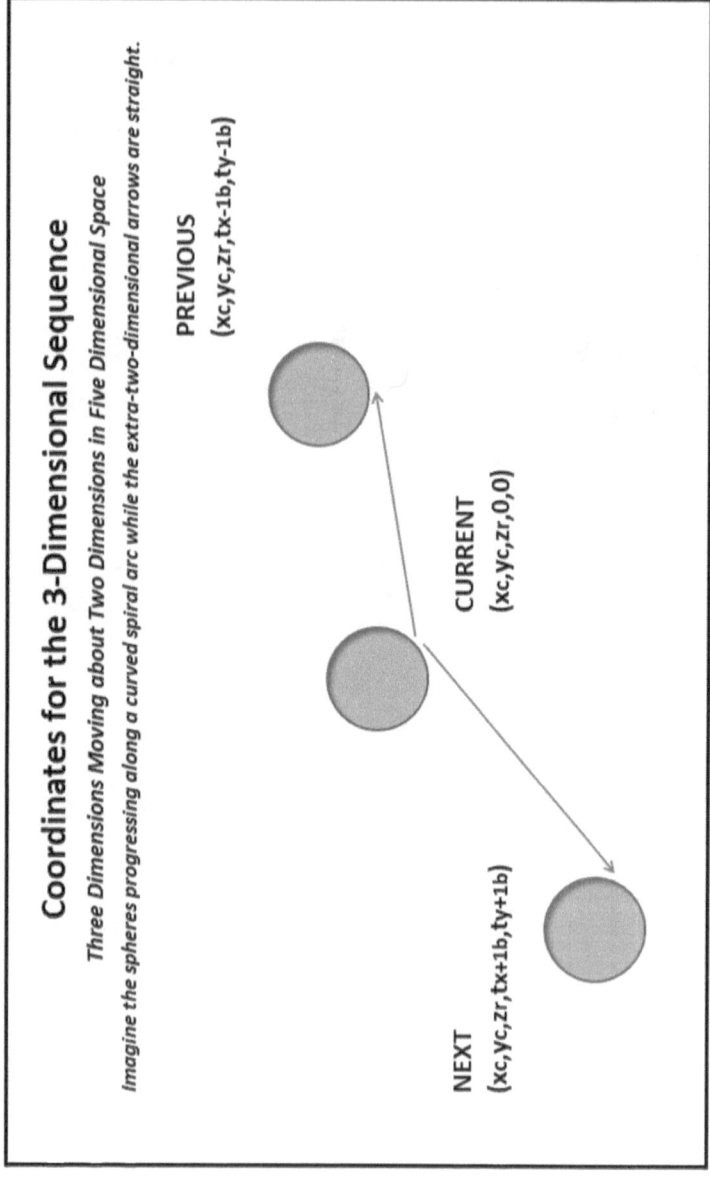

Coordinates for the 3-Dimensional Sequence

Three Dimensions Moving about Two Dimensions in Five Dimensional Space

Imagine the spheres progressing along a curved spiral arc while the extra-two-dimensional arrows are straight.

PREVIOUS
(xc,yc,zr,tx-1b,ty-1b)

CURRENT
(xc,yc,zr,0,0)

NEXT
(xc,yc,zr,tx+1b,ty+1b)

Diagram 7. Figurative Tangent Model for 3 + 2 = 5

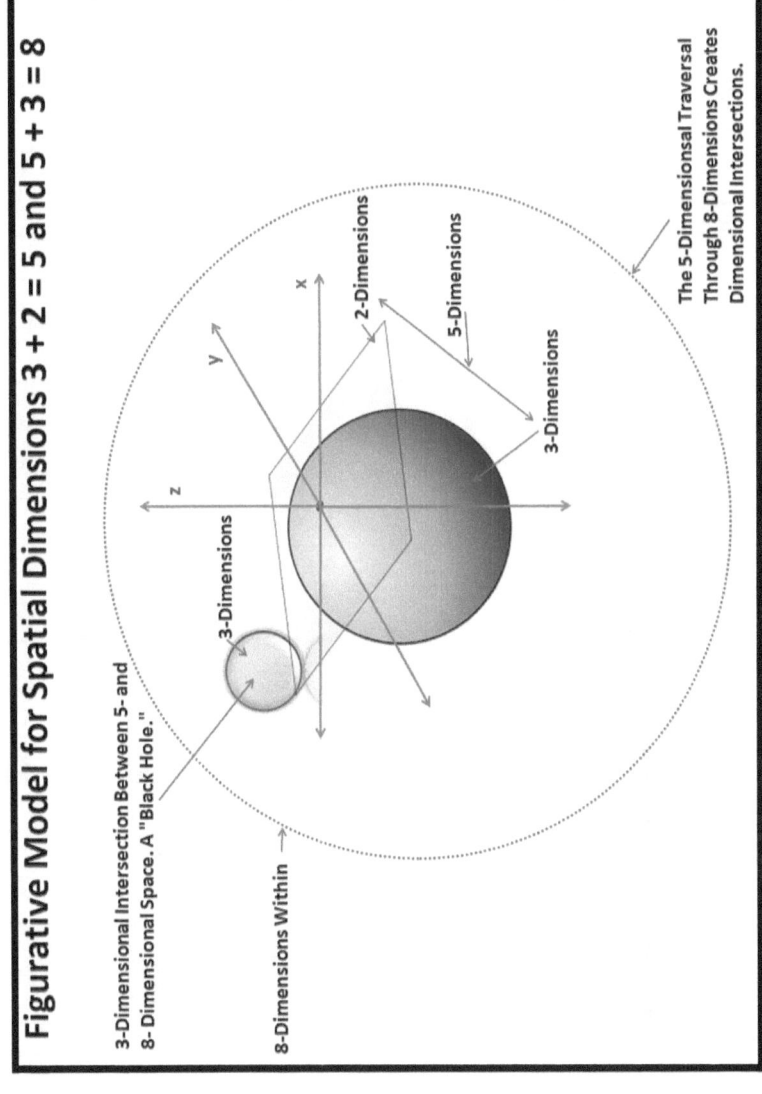

Figurative Model for Spatial Dimensions 3 + 2 = 5 and 5 + 3 = 8

3-Dimensional Intersection Between 5- and
8- Dimensional Space. A "Black Hole."

3-Dimensions

8-Dimensions Within

x

y

z

2-Dimensions

5-Dimensions

3-Dimensions

The 5-Dimensionsal Traversal
Through 8-Dimensions Creates
Dimensional Intersections.

Diagram 8. 5D Computer Schematic Model

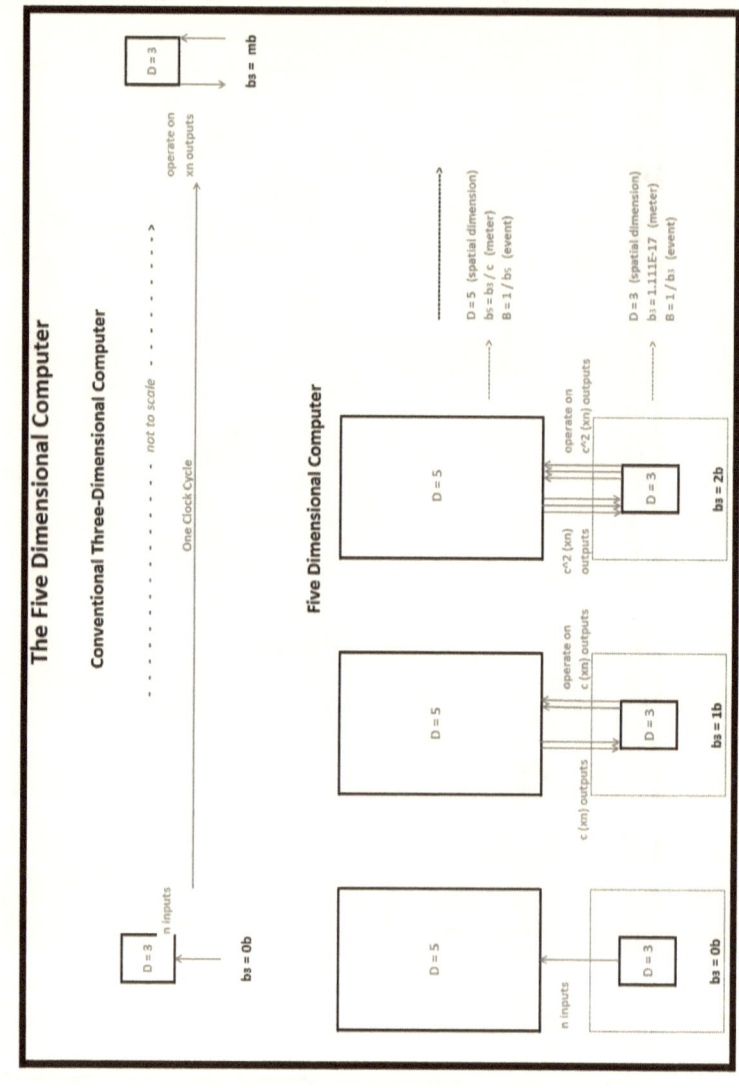

The Five Dimensional Computer

Conventional Three-Dimensional Computer

D = 3

bs = 0b

n inputs

· · · · · · · · · · · · · · · not to scale · · · · · · · · · >

One Clock Cycle

operate on xn outputs

D = 3

bs = mb

Five Dimensional Computer

D = 5

n inputs

D = 3

bs = 0b

D = 5

c (xn) outputs

operate on c (xn) outputs

D = 3

bs = 1b

D = 5

c^2 (xn) outputs

operate on c^2 (xn) outputs

D = 3

bs = 2b

D = 5 (spatial dimension)
bs = bs / c (meter)
B = 1 / bs (event)

D = 3 (spatial dimension)
bs = 1.111E-17 (meter)
B = 1 / bs (event)

Diagram 9. Geometric Considerations for 5D PC

Top / Bottom View - Positive Charge Utilized Material

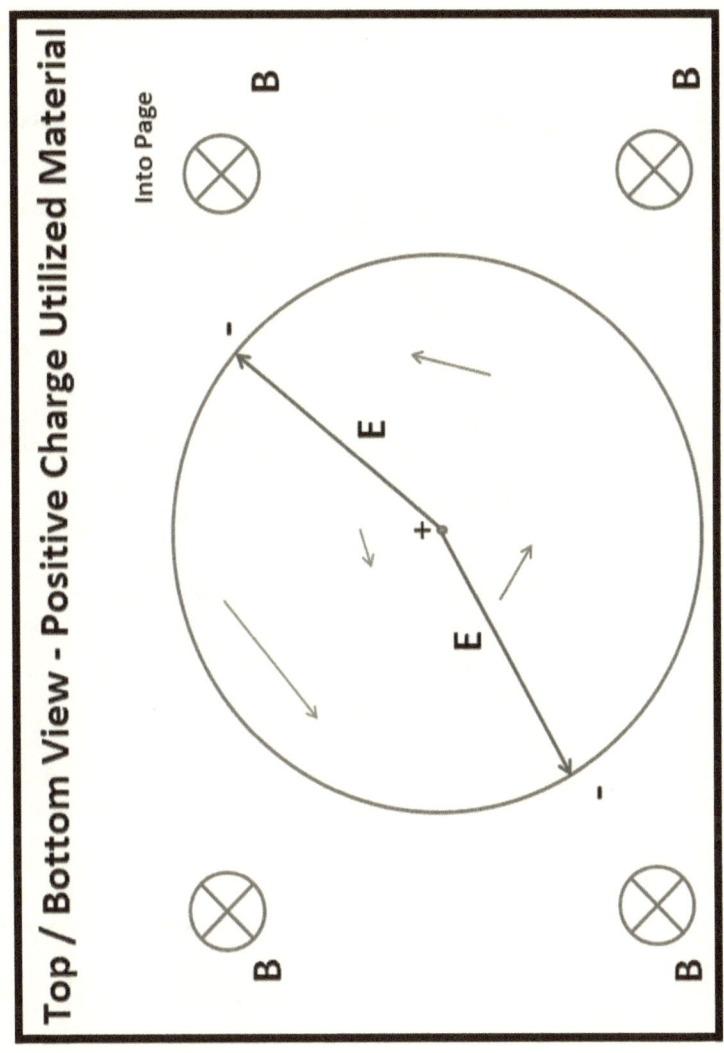

Into Page

B

B

B

B

E

E

+

-

-

Diagram 10. Topical Geometry Considerations for 5D PC

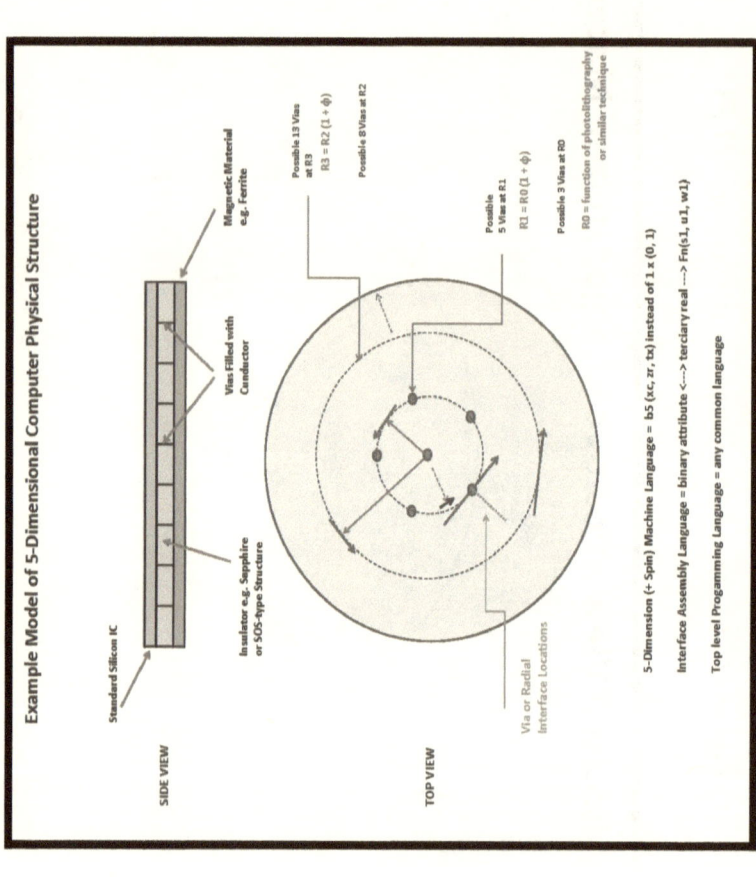

Example Model of 5-Dimensional Computer Physical Structure

SIDE VIEW

Standard Silicon IC

Vias Filled with Conductor

Magnetic Material e.g. Ferrite

Insulator e.g. Sapphire or SOS-type Structure

TOP VIEW

Via or Radial Interface Locations

Possible 13 Vias at R3

$R3 = R2 (1 + \phi)$

Possible 8 Vias at R2

Possible 5 Vias at R1

$R1 = R0 (1 + \phi)$

Possible 3 Vias at R0

$R0$ = function of photolithography or similar technique

5-Dimension (+ Spin) Machine Language = b5 (xc, zr, tx) instead of 1 x (0, 1)

Interface Assembly Language = binary attribute <---> terciary real ---> Fn(s1, u1, w1)

Top level Progamming Language = any common language

66

Diagram 11. Example 5D PC Logic Application

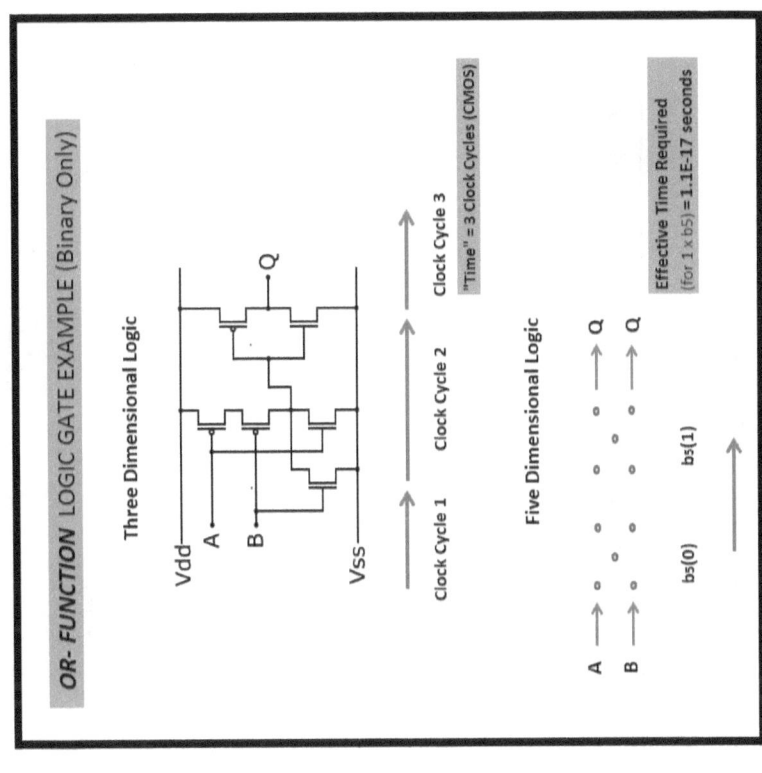

OR- FUNCTION LOGIC GATE EXAMPLE (Binary Only)

Three Dimensional Logic

Vdd
A
B
Q
Vss

Clock Cycle 1   Clock Cycle 2   Clock Cycle 3

"Time" = 3 Clock Cycles (CMOS)

Five Dimensional Logic

A → o o o o o → Q
B → o o o o o → Q

bs(0)   bs(1)

Effective Time Required
(for 1 x b5) = 1.1E-17 seconds

Diagram 12. The Meaning of Spin

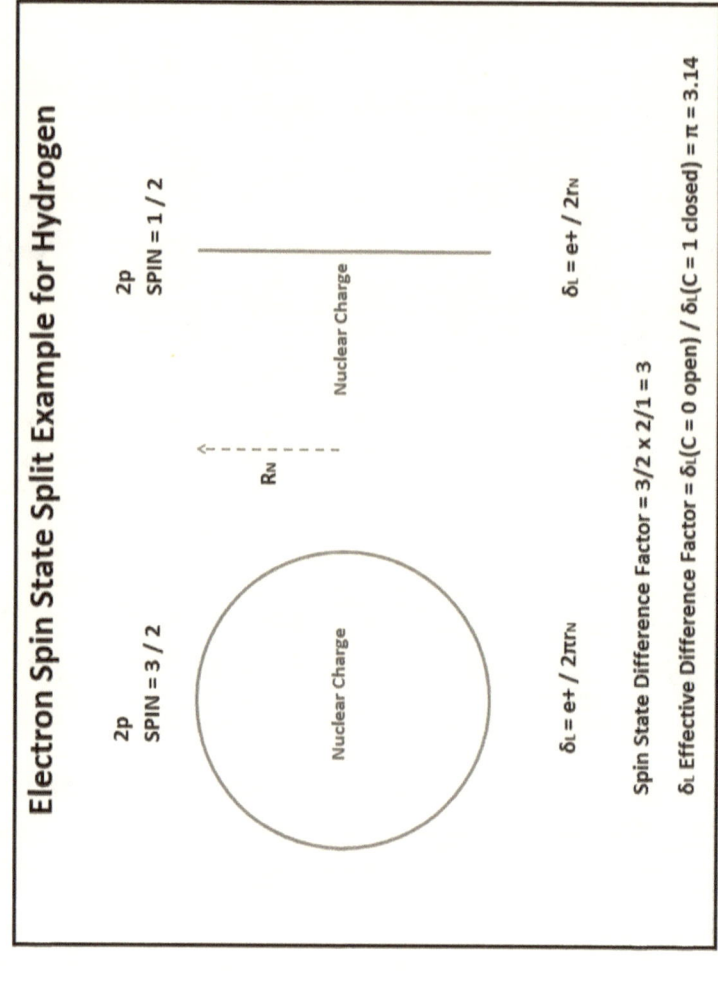

Electron Spin State Split Example for Hydrogen

2p
SPIN = 3 / 2

2p
SPIN = 1 / 2

Nuclear Charge

Nuclear Charge

R_N

$\delta_L = e+ / 2\pi r_N$

$\delta_L = e+ / 2r_N$

Spin State Difference Factor = 3/2 x 2/1 = 3

δ_L Effective Difference Factor = δ_L(C = 0 open) / δ_L(C = 1 closed) = π = 3.14

Diagram 13. Transition Model in Higher Dimension

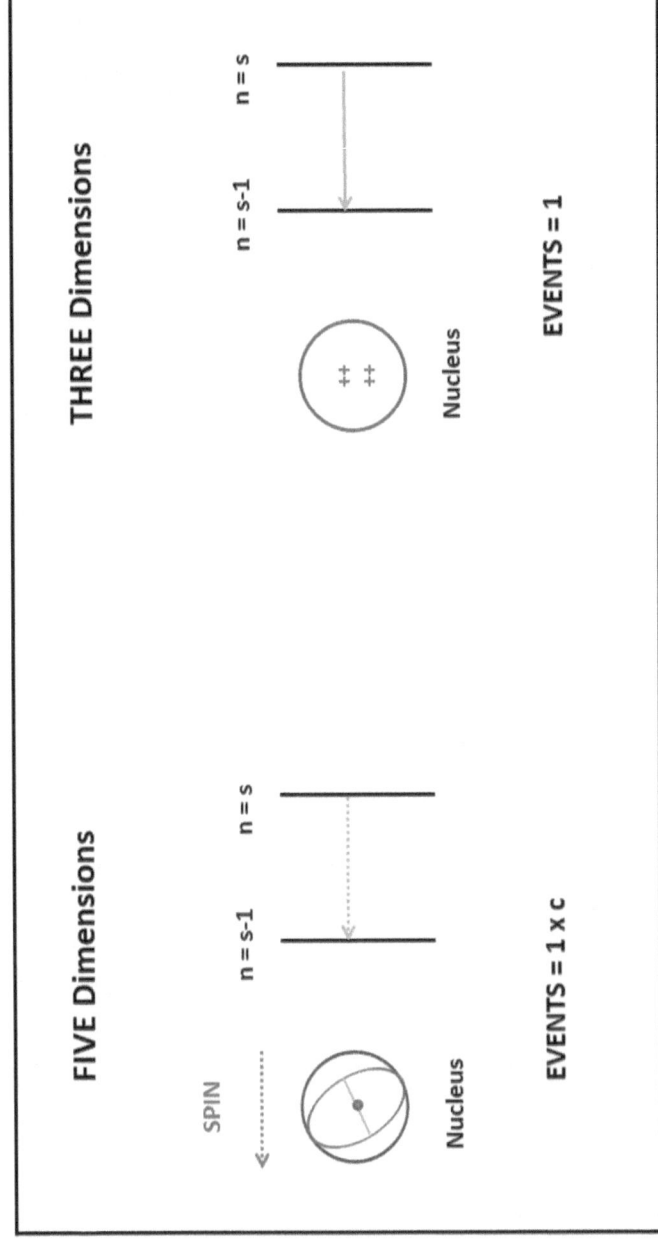

Diagram 14. Model for Cold Fusion

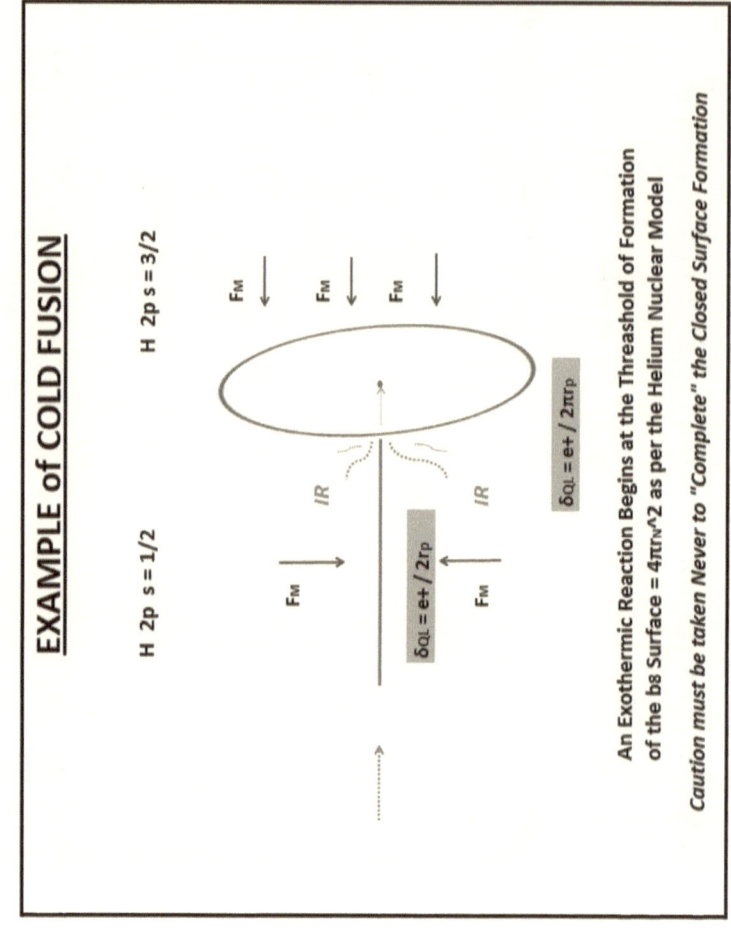

EXAMPLE of COLD FUSION

H 2p s = 1/2   H 2p s = 3/2

F_M

F_M

F_M

F_M

IR

F_M

IR

$\delta_{QL} = e^+ / 2r_P$

$\delta_{QL} = e^+ / 2\pi r_P$

An Exothermic Reaction Begins at the Threashold of Formation of the b8 Surface = $4\pi r_N{}^2$ as per the Helium Nuclear Model

Caution must be taken Never to "Complete" the Closed Surface Formation

Diagram 15. Flat Earth Map

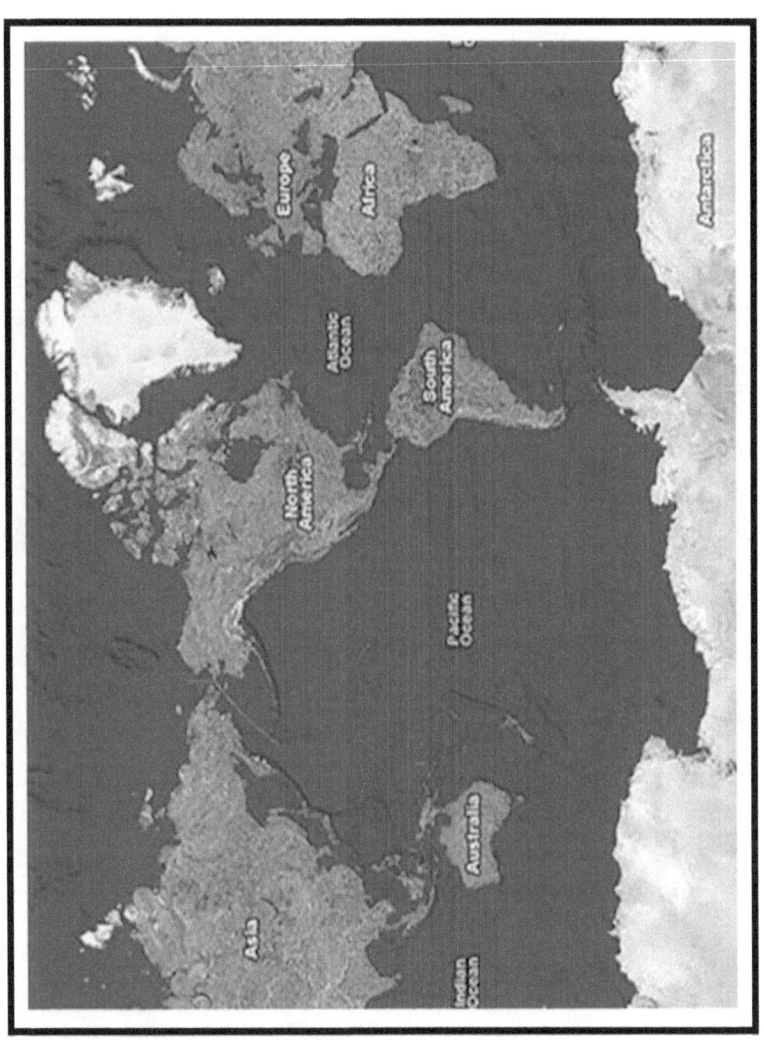

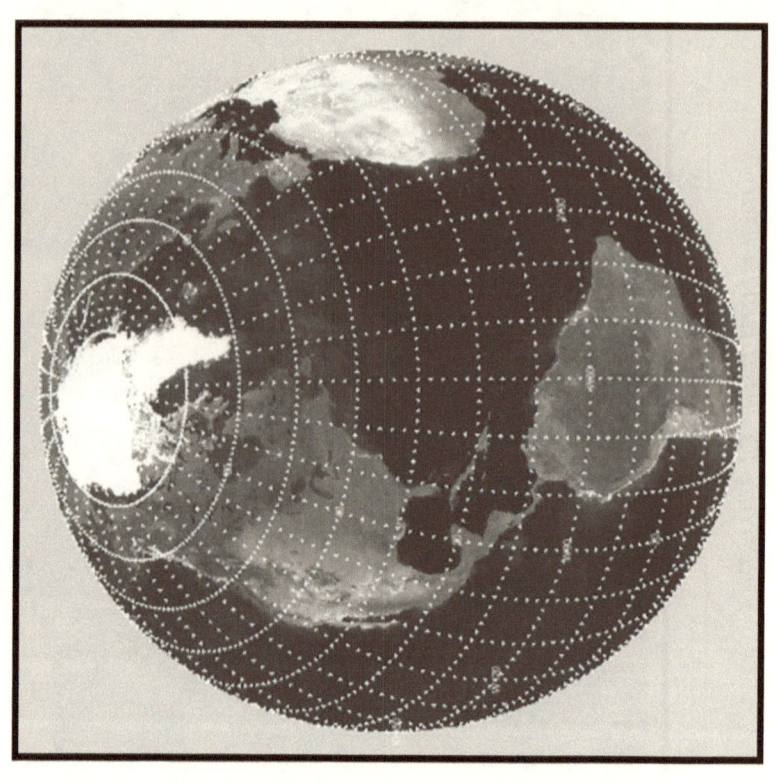

Diagram 16. Spherical Earth Map

Diagram 17. 5D Computer Bit-Map

Example Bit Map for Compiler

clock	b3	b5	#	word 1 bit 1	bit 2	bit 3	bit 4	bit 5	bit 6	bit 7	bit 8	word 2 bit 1	bit 2	bit 3	bit 4	bit 5	bit 6	bit 7	bit 8
0sec				1	0	0	0	1	0	0	1	1	0	0	0	1	0	0	0
	1		1	s0	s0	s0	s0	s0	s0	s0	s1	s0	s0	s0	s0	s0	s0	s0	s1
			2	s1	s0	s0	s0	s0	s0	s1	s2	s1	s0	s0	s0	s0	s0	s1	s2
			3	s2	s1	s0	s0	s0	s1	s2	s3	s2	s1	s0	s0	s0	s1	s2	s3
			4	s3	s2	s1	s0	s1	s2	s3	s4	s3	s2	s1	s0	s1	s2	s3	s4
			5	s4	s3	s2	s1	s2	s3	s4	s5	s4	s3	s2	s1	s2	s3	s4	s5
			6	s5	s4	s3	s2	s3	s4	s5	s6	s5	s4	s3	s2	s3	s4	s5	s6
			7	s6	s5	s4	s3	s4	s5	s6	s7	s6	s5	s4	s3	s4	s5	s6	s7
			8	s7	s6	s5	s4					s7							
			9																
			10																
			--																
			n	sn-0	sn-1	sn-2	sn-3	sn-2	sn-1	sn-0		sn-1	sn-2	sn-3	sn-3	sn-2	sn-1	sn-2	sn-3
			--																
			c	sc-0	sc-1	sc-2	sc-3	sc-2	sc-1	sc-0		sc-1	sc-2	sc-3	sc-3	sc-2	sc-1	sc-2	sc-3
3.3E-09sec	2			A-0	A-1	A-2	A-3	A-2	A-1	A-0	A-1	A-1	A-2	A-3	A-3	A-2	A-1	A-2	A-3

where A(x) = Compiler Operator $A(x) = A(0,1)$

73

Diagram 18. Example Nuclear Spin Positions

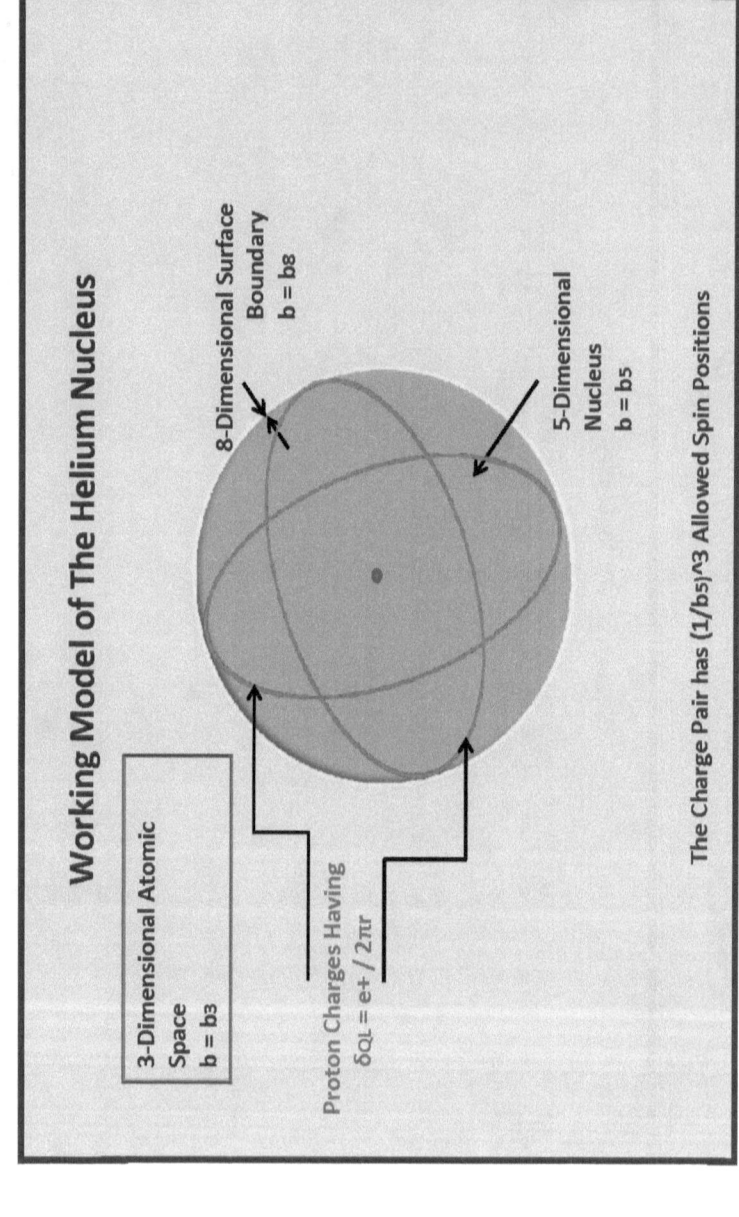

Working Model of The Helium Nucleus

3-Dimensional Atomic
Space
$b = b_3$

8-Dimensional Surface
Boundary
$b = b_8$

5-Dimensional
Nucleus
$b = b_5$

Proton Charges Having
$\delta_{QL} = e+ / 2\pi r$

The Charge Pair has $(1/b_5)^3$ Allowed Spin Positions

Diagram 19. $E = hv = h(c/\lambda)$

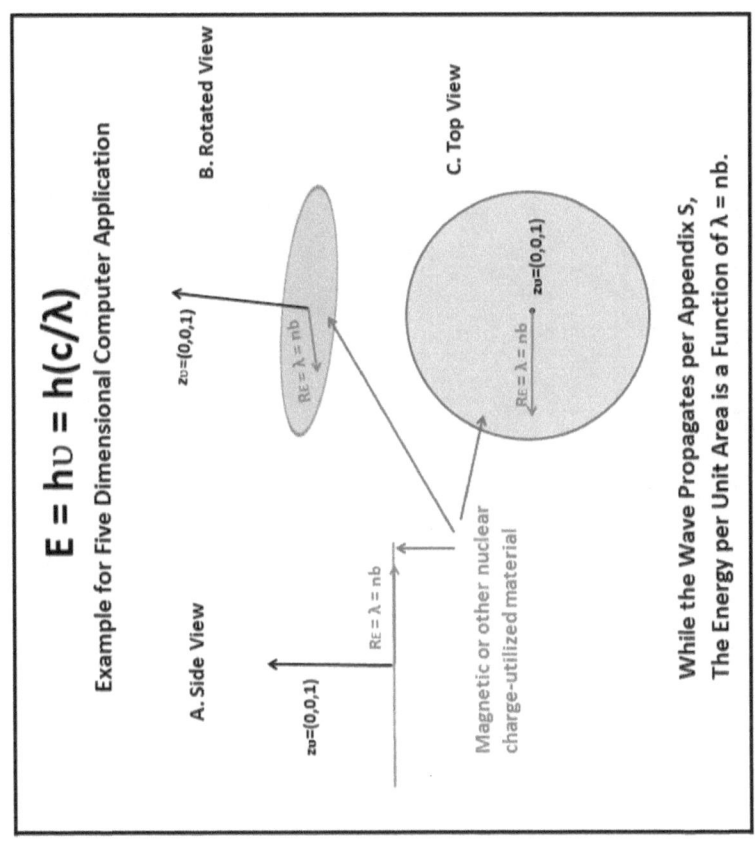

$E = h\upsilon = h(c/\lambda)$

Example for Five Dimensional Computer Application

A. Side View

$z_\upsilon = (0,0,1)$

$R_E = \lambda = nb$

Magnetic or other nuclear
charge-utilized material

B. Rotated View

$z_\upsilon = (0,0,1)$

$R_E = \lambda = nb$

C. Top View

$R_E = \lambda = nb$ $z_\upsilon = (0,0,1)$

**While the Wave Propagates per Appendix S,
The Energy per Unit Area is a Function of $\lambda = nb$.**

Diagram 20. Bit Placement (5DPC)

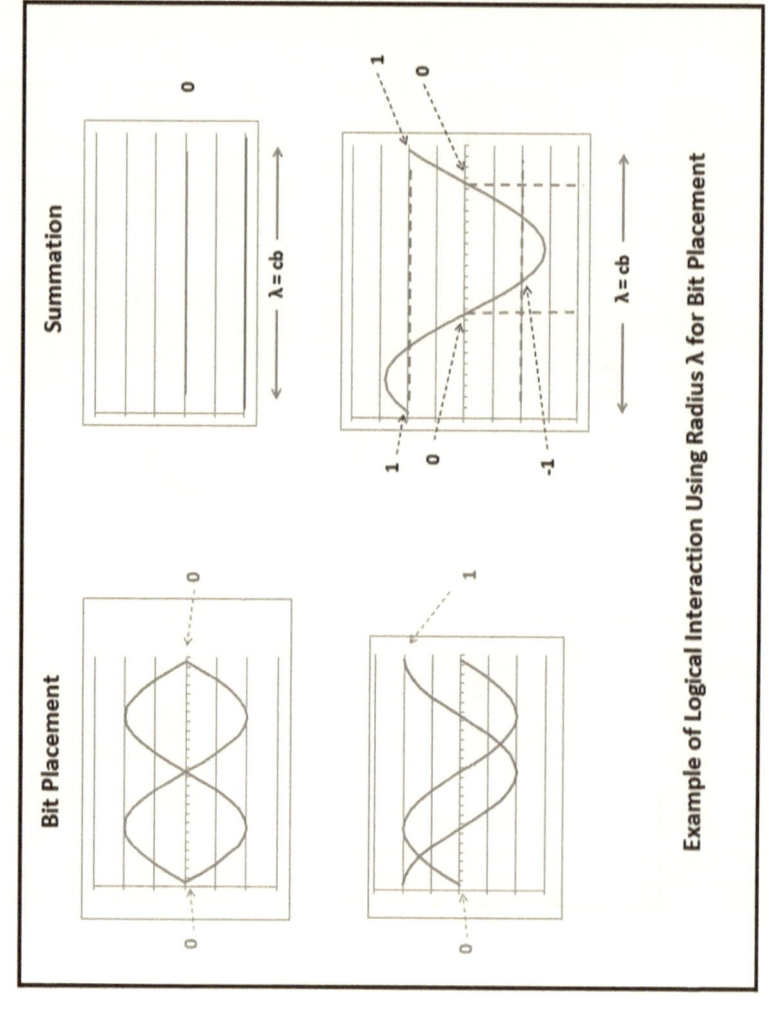

Example of Logical Interaction Using Radius λ for Bit Placement

Diagram 21. Dimensional Gravitational Spatial Progression

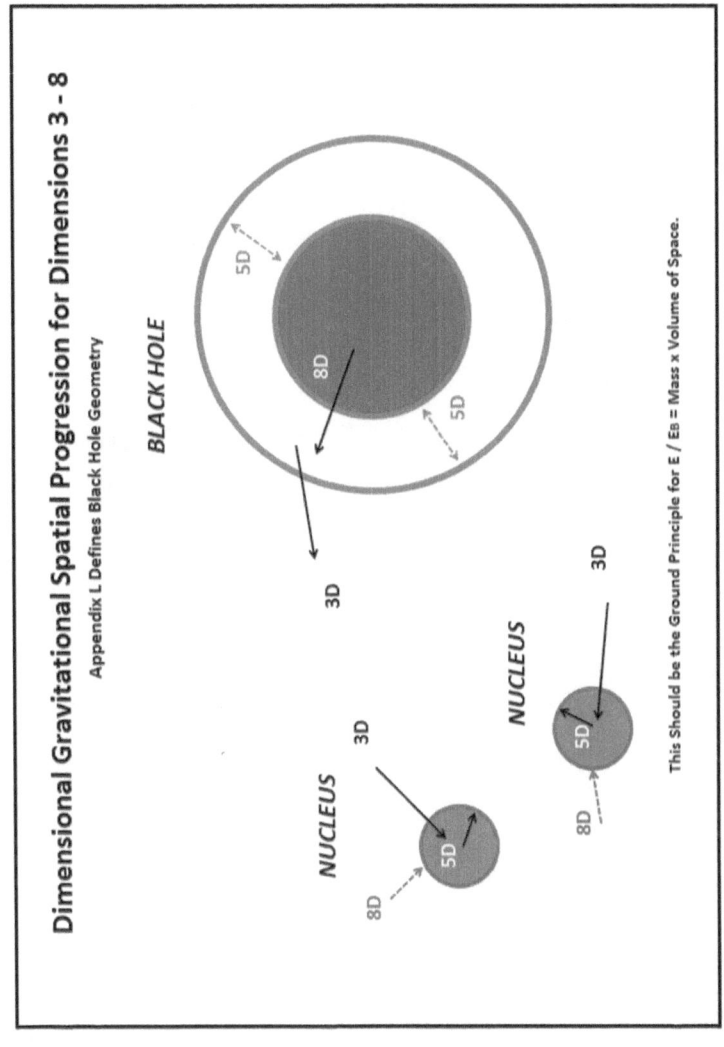

Dimensional Gravitational Spatial Progression for Dimensions 3 - 8

Appendix L Defines Black Hole Geometry

BLACK HOLE

5D

5D

8D

3D

NUCLEUS

3D

NUCLEUS

5D

8D

5D

3D

8D

This Should be the Ground Principle for E / E$_B$ = Mass x Volume of Space.

XII. *Appendices and Technical Appendices*

While the dimensional model t=cB is unequivocally proved in the technical Appendix W, it is recommended for all readers to review the geometries shown in the short Appendix X. As always, ignore the geometries if they do not seem intuitive.

Introduction to Technical Appendices

$E = h\nu$ (Revisited)

Information at the Rate c^2

Appendix U Further Description of the 5-Dimensional Computer

Appendix V One Figurative "Type-Model" for Cold Fusion

INTRODUCTION

$E = h\nu$ must be defined without the concept of time even though we can get around the old "time" model by using $\nu = c / \lambda$ where c is numerical and not a "velocity."

E = hν (REVISITED)

$E = h\nu$ consists of three factors:

E changes, h changes per Appendix D,

$\nu = c / \lambda$ changes per Appendix E.

While at the same "time,"

E changes, h changes per Appendix D,

$\nu = c / \lambda$ is fixed per Appendix S. $h = bE_B\kappa$ can we written as $h = E_s\kappa / mb^2$ where

$E_s\kappa$ represents the intrinsic energy of space per the main transcript, and $h = h(E_B)$ represents a "slope" as a function of spatial energy (location with respect to mass.)

In the absence of time t, the vector expression per frame becomes:

$E_f = h\nu = h(1/\lambda) \times (0, 0, 1)$ (energy per orthogonal unit boundary area) where

$\lambda = (x^2 + y^2)^{1/2} = R_E = nb$ and satisfies all conditions.

INFORMATION AT THE RATE c^2

As one example, in Diagrams 10 and 19, imagine the magnetic material is the non-magnetic and non-solid-state material hydrogen (H.)

In that figurative case, the 2p spin split energy has a wavelength of 2.7 cm.

Imagine a radial closed geometry (in our forced two-dimensional surface) with $R = 2.7$ cm.

Now imagine concentric closed geometries each with radius 1, 2, 3, and 4x the value 2.7cm / 4 respectively.

As shown directly above and per Diagram 3, then we have established logical conditions at each of the 4 radial locations and have done so faster than the three-dimensional (electronic) rate of progression, i.e. > c.

Initialized radial values, for example, would be 0, 1, 0,-1, and back to 0 respectively from smallest to largest radius.

These values are large enough to be read using modern embedded transducers.

APPENDIX U

CONSIDERATIONS FOR THE FIVE DIMENSIONAL COMPUTER MODEL

As shown in the main text and diagrams, logical interactions can be accomplished in-between standard electronic "bits" so that computational, operational, and processing actions can take place at a much higher rate than standard electron-science allows. See Diagram 20.

The method, while somewhat of a misnomer, can be called "magnetic" i.e. using the positive nuclear (proton) charges to extremely enhance the rate of information exchange.

It has clearly been shown that the atomic nucleus has "informational" positions using very low energies to greatly exceed standard electronic processing capability (where electronic is defined as depending solely on the electron and ignoring the positive nuclear charge.) See Diagram 18.

Existing methods for using positive nuclear charge have so-far been confined to media and magnetic resonance and allegedly have processing speeds even lower than standard electronics.

Per the diagrams, the "electronic" route follows a spatial pattern. The "nuclear charge" route is capable of the number c (Joule system 3E+08) within one "single" electronic motion.

A simplistic interface is a fiber-optic means for initiating energy states that progress in an orderly forward progression, upon initialization, using low thermal energies in the range of "spin-split" atomic energies.

As a simple example, the Hydrogen "fine" spin-split energy is on the order of an 2.7cm IR wave.

The interface "vias" (figurative) must be along the spin-progression patterns.

Per the diagrams, the electronic physical radius $R(\varphi)$ increases at an effective rate $(1/c)$ x the rate of nuclear-charge progression along the curved surface x_c.

The spin relation of nuclear charges and the exact locations of the relative spins are geometrically critical.

Initialized bits are received or read by geometric location as follows:

First: The exact locations of the initializing bits are important to be read.

Second: The exact radial locations, using the geometric growth rate φ and the wavelength λ, are important to be read.

Third: The exact locations along the "curved" spiral having growth rate φ are important to be read.

READING AND WRITING

Writing is straight-forward and requires enough energy to provide the "spin-distinguished" energy (e.g. but not limited to the H 2.7 cm IR spin-split wavelength) per unit area (see example drawings) in order to populate the nearby three and five dimensional bits along their respective distances.

Reading, depending on the material used, could be recognition of small energies, on the order of spin-split energies, using enough surface area to be interfaced and inserted in standard three-dimensional electronic logic.

INTERFACE

Interface between the electronic (large and cumbersome) and nuclear-spin (finely exact) energies can be accomplished by many methods including fiber optics, voltage across a thin film and/or a thermocouple film, and by standard magnetic means.

APPENDIX V

ONE FIGURATIVE EXAMPLE METHOD FOR REACTING NUCLEI

Consider two separate containers of Hydrogen gas (H_2) that are sealed from reacting gasses such as oxygen.

Each container has pressure and temperature controls.

The containers are connected by an effective Venturi column of controlled dimensions.

An electrical field **E** and a magnetic field **B** are controlled at required distances in order to produce uniform force fields across the dimensions of the containers and Venturi column. Some values may be 0 (zero.)

Through controls, a low energy (IR) wave should not be present in the "upper" bounds of the container(s) adjacent to an operational surface.

The upper container then contains H_2 where the bonding valence energy 2p should have the value 2p spin 1/2 and the lower container contains random H_2 gas that has approximately 25% spin 1/2, 25% spin 3/2, and 50% spin 1/1.

During the force of entrance to the "cold" chamber, the molecules quickly achieve the same planar position as the existing "cold" molecules due to the **E** and **B** force fields there.

The entrance Hydrogen meets the "cold" Hydrogen at the surface of interaction where the "spin" has mostly been fixed.

The collision interface should statistically produce exothermic reactions when and where a fraction of the 25% spin = 3/2 Hydrogen encounters the near 100% spin = 1/2 Hydrogen of the fixed chamber.

The resulting thermal energy can be captured about the surface-of-reaction. See Diagram 14.

APPENDIX W

ATTRIBUTES

We suggest the "charge-force" is three-dimensional nomenclature for a spatial dimensional force or push-pull "oscillation." For example, from observation:

Mass pulls mass weakly (gravity) in the same dimension.

Mass (D=N) and mass (D=N +/- 1) pull each other strongly.

We have shown energy is a function of dimensional spatial gradient (Appendix Q) and that E / E_B has units of mass x volume (Appendix E.)

We have shown time $t = cB = (c/b)$ per the main manuscript.

We suggest there are two and only two physical realities:

Space
Mass

It is possible there are no other physically real attributes.

In that case, the "charge-force" should be directly related to the rate of sequential spatial progression c.

We can write the unified gravitational force in three dimensions as

$$F = Km_1m_2 / r^2 \text{ (Nwt) or}$$

$$F_{m2} = K / r^2 \text{ (Nwt-Kg}^{-2}\text{) where}$$

$$K(D) = K(\Delta D) + G$$

and where

G = the known gravitational constant,

$\Delta D = 1$ for the change D=5 to D=3 and $\Delta D=0$ (zero) for no dimensional change,

and K can be derived as follows:

(for example, using the Bohr[4] model for Hydrogen states n=1 and n=2,)

$$-10.2eV = Km_pm_E (1 / r_2 - 1 / r_1).$$

Then

$K = 7.57E+28$ (Nwt- kg $^{-2}$ meter ^{2})

and represents a large force that interacts between D=3 and D=5.

As a constructive exercise,

using time-based physics, we could then write:

$F = ma$ where

$a = $ meters / second^(D-1).

Per Appendices C and L:

For 3-dimensional space, D-1 = 2.
For 5-dimensional space, D-1 = 3.
For 8-dimensional space, D-1 = 5.

Continuing with the Bohr model for Hydrogen state n=2 to n=1, then

$F(adj) = m_E (4r_B - r_B) / t^5$. Then,

$t^5 = 3m_E r_B / F = (3m_E r_B / Km_E m_P) \int r^2 \, dr$ (from r_B to $4r_B$), and

$t = (3.90E\text{-}42)^{\wedge}1/5 = 5.23E\text{-}09$ sec.

For the example transition, then

$t = (1 + \kappa) / c$

within the mean factor (0.0029) or 0.29% where

κ finely defines Planck's constant[5] (h)

$h = b\, E_B\, \kappa$

per Appendix D (attached.)

The dimensional factor κ then has a clearer meaning.

Per Appendix D,

$\kappa = .08^{\wedge}5/2 / e^{\wedge}3/5 = \gamma' / \varphi'$.

γ' represents an "extra" factor of distance per dimension (one-dimensional-line) per Appendices A through D and also Appendix Q, while

φ' represents the local dimensional change for space itself per Appendix C, i.e. the rate 0.600 instead of the universal limit 0.618.

In this example, using the transition expression from above

$$t = (1 + \kappa) / c = 1 / c + \kappa / c,$$

units of γ' are represented as "sec-dimension^{-1}" while

units of φ' are "dimension^{-1}"

as we are using the value c numerically (number of events in one second instead of a "velocity.")

Then κ becomes a dimensional adjustment for the concept of time in the same way it adjusts the value h,

$$h = b \, E_B \, \kappa \quad \text{per Appendix D}$$

and also defines the value of transitional n,

$$n = \kappa c \quad \text{per Appendix E.}$$

In fact, we notice "time t" in this case is the sum of a three dimensional term and a five dimensional term:

$$t = 1 / c + n_0 / c^2 \quad \text{where}$$

$$n_0 = \kappa c.$$

This suggests the form of an infinite geometric series.

We recognize $1 / c$ as the smallest increment of three dimensional time (using the "time" model.)

We recognize n_0 / c^2 as the increment of five dimensional "time" required to define the quantum energy at the intersection of bound E and unbound E,

where $E / E_B = 1$ (per Appendix E.)

This should be the principle for so-called time dilation since we would then perceive time to "elongate" as an inverse function of E_B (and as a strong function of higher dimensional space.)

Verification of Calculations

As verification of our thinking, we can equivalently use the derivation from Appendix Q to prove the constant K using an alternate and independently derived energy.

Instead of the quantum energy -10.2eV for Hydrogen states 1 and 2, we should achieve the same result using the force (gradient) **Grad E_B** exactly as in Appendix Q.

Then Grad $E_B = c^3$ from three to five dimensional space (at the nuclear surface.)

In Appendix Q, we calculated the nuclear radius of "zero-neutron" Helium as 3.07E-14 meters. (Neutrons are not discussed in this appendix.)

Postulating for now, that the H nucleus should be close to the same size, then we can approximately write

$E = c^3 \, r_N = 8.29E+11$ (J) for the H nucleus. Using the identical Bohr model from above, it follows directly that

$E \, (1/r_B - 1/4r_B) = m_E \, (a_{rB} - a_{4rB})$. Then,

$\Delta a = 1.30E+52$ and

$\Delta F = m_E \Delta a = 1.18E+22$ (Nwt).

As a proof, we need to relate dimensional force F from Appendix Q and the uniform gravitational constant K:

Set $1.18E+22 = \Delta r_N \, K \, m_p \, (3/4) \, r_B^{-2}$.

Then we must correct our approximation of the H radius r_N:

$\Delta r_N = 4.0$ and

$r_N(He) \, / \, r_N(H) = Z^2$ from the old science.

These should represent the nature of all things large and small alike.

Similarly, we now see the meaning of empirical Z(eff) from the older quantum science:

In Appendix Q, we showed the actual He r_N to be $2^{(1/6)}$ larger using 4 amu (stable) He instead of a proposed 2 amu He.

Then considering the charge attribute alone (incorrect,) we would observe

$E_N(old)/A_N(old) = 2^{(1/6)} E_N(actual)/A_N(actual)$ and

$Z(empirical)/Z(actual) = 1/(2^{1/6})$ for He.

We would then perceive,

$Z(old) = 2/(2^{1/6}) = 1.782$

and the old model would need to "adjust" Z_e to fit an Hamiltonian[6] electron operator that was in fact using an incorrect model for the nuclear energy itself.

(Unknown using the old model, the nuclear energy was in fact smaller.)

The actual Z_e adjustment-to-fit would then need to vary depending on the given electron radius from the nucleus.

The model t=cB precisely corrects the nuclear and atomic approximations from the continuous-time model.

APPENDIX X

POSSIBLE KERNEL

In mathematics, a kernel is the smallest constituent or basis upon which everything else is built or advances.

For the low energy communication of hydrocarbons (perhaps the basis for the growth of life itself,) a suggested kernel is CH_3:

At once, the operational group is geometrically fixed, capable of exchanging hydrogen spin state energies within itself and among other operational kernels, and is prolific.

Related diagrams follow below.

Initialization of energies and continuity of the unaltered low energy wave within any actual kernal or kernel group should be critical to maintain natural sequences. As an example, radioactive charge energies are known to cause morphed growth sequences.

$$R-CH_3$$

$$R-\overset{\displaystyle H}{\underset{\displaystyle H}{C}}-H$$

"PYRAMID - LIKE"
(TETRAHEDRON)

KERNEL DIAGRAM

With appropriate mathematical transformation, the following kernel diagram is clear. This is not dissimilar to the concept shown in Diagram 7:

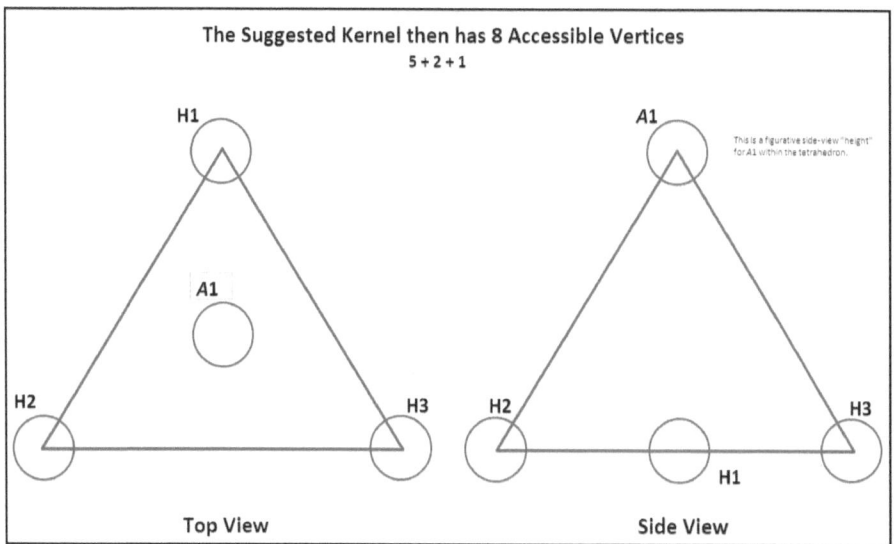

An Aside:

Imagine you are the pilot of a fighter-jet.

Your reaction-based computer operates at less than the rate c.

Your opponent's reaction-based computer operates at the rate c^2.

In this case, you may be the "losing dog" in the dog fight.

APPENDIX Y

NEUTRON MASS

As proton mass is five dimensional, we propose neutron mass is eight dimensional. Then,

$$m_N = \delta_N \times 4\pi r_N^2 = m_P.$$

Per Appendix W, while

m_N, m_P, and m_E all react weakly proportional to G, the pairs

m_E, m_P and m_P, m_N then react strongly proportional to K.

Two neutrons in eight-dimensional space should experience two-dimensional masses weakly acted upon through three-dimensional space,

$$F = Gm_N^2 / r^5.$$

Mass surfaces should then be drawn to "coincide" as a result of this force.

The distance r between the two radii cannot be zero and cannot be smaller than b_{13} per Appendix L.

While G is small, F is large.

A proton and neutron in the same atomic nucleus should interact as shown in Appendix W through the "strong" constant K using the denominator $r_N{}^3$.

This strong force should create quantum energy states within the nucleus related to b_5 and its own **GradE$_B$**, and so on.

As shown in Appendices Q and R, the neutron mass in the nuclear three-dimensional intersection between 5 and 8 dimensional space must obey the atomic boundary condition mass x volume.

While the nucleus achieves a lower energy state through the addition of neutron mass, the atomic mass x volume remains bounded.

In three dimensions, the masses m_E and m_P are very different.

In five dimensions, the masses m_P and m_N are very different based on the dimensional difference in E_B.

In three dimensions, the difference between m_P and m_N cannot be distinguished.

We suggest that most of the mass that "tips" a three-dimensional measurement scale is five and eight dimensional mass that resides within a 3-dimensional intersection between 5 and 8 dimensional space.

APPENDIX Z

FURTHER EXAMINING SPIN AND E = hν

From $t = cB$, we can derive the following per the main manuscript:

$v = c / \lambda$ cycles per second.

For the H $2p_{3/2}$ – H $2p_{1/2}$ spin split transition, $\lambda = .027$ meters, then

$v = c / \lambda = b$ E+27 cycles per second, or

$v = b$ E+27 $/ (cB)$ cycles per event. It follows,

$v = (b^2 / c)$ E+27 cycles for the spin split transition, or

$v = (b^2 / c) (c^3 / \lambda)$ cycles, and

$v = (cb)^2 / \lambda.$

Equivalently, we can write:

$v = c^2 / \lambda B^2.$ This has the form of a quantum expression.

We can define the maximum possible v,

$$v = c^2 / \lambda.$$

We can quickly show:

$v = c^2/ \lambda = (b_3 c^0 \kappa^6)^{-1} = (b_5 c^1 \kappa^6)^{-1} = (b_8 c^3 \kappa^6)^{-1}$. Then

$n = \lambda v \kappa^6$ where $\lambda = n b_3$ per Appendix E.

In this case, $\kappa = \gamma'/\varphi'$ and represents the 3-to-5 dimensional spatial transition per appendices D, E, and W.

Per Appendix Q, a one-dimensional radius in 5-dimensional space obeys a 6th root law.

We do not intend to generalize the spin split calculation and expression for higher dimensional v (meter^{-1}) in the present scope.

CLOSED SPATIAL TRAVEL:

Consider the mass-matter that has the smallest possible natural (not externally excited) |energy| 2-D surface within 3-dimensional space.

This should be the surface having the radius r' , where r' represents the 2-D cross-section radius from the single proton nuclear center of an H atom to the ground electron state for the same H atom.

Next, consider the spatial traverse for the single 3-D H (hydrogen) atom through 5-dimensional space:

From the 5-dimensional view, 3-D geometry is closed (curvature=1) in 5-dimensions. Then the 3-dimensional space must traverse a closed 5-D "circle" through the additional two (of the total five) dimensions in 5-D.

Except a 5-D "circle" is not the same as a 2-D circle that can be simply integrated into a closed surface and subsequently the surface into a "volume." While straight-forward, the next volume (5-D) integration involves two dimensions expanding into five.

In 2-D, the relationship for a closed-distance is $d_{C2} = 2\pi^1 r$ because the spatial change is only to-and-from a delta of one dimension. It directly follows, in 5-D the relationship for a closed-distance is $d_{C5} = 2\pi^3 r$.

The minimum allowed distance for the two dimensional surface to progress along the direction (0,0,"$\mathbf{v}/v$") to "return to where it started" or to return "back onto itself" (close itself) is cb_3.

The increase in one-dimensional size (r) from 3 to 5 dimensions is .08^(5/3) per Appendices C and Q.

Then,

$2\pi^3 r(1+.08^{5/3}) = cb_3$ *where*

$c = 2.998E+08$ *events and*
$b_3 = 1.111E-17$ *meter-event^{-1}.*

Then,

$r = cb_3 / 2\pi^3(1+.08^{5/3}) = 5.29E-11$ *meters* $= r_B$. *In other words,*

$r = r' = r_B$ *the exact Bohr^{4}radius.*

To generalize:

Let $r_5 = r_3 (1+.08^{5/3})$.

For multiple mass-pair (m_P, m_E),

$$r_s = cz^2 b_3 / 2\pi^3,$$

and as a function of energy,

$$r_s = cn\lambda / 2\pi^3,$$

where r_s is a function of the number z of (m_P, m_E) pair and is also a function of E/E_B.

To continue with the H example:

In 3D, the radius r is not fixed and the 3D geometry is not closed.

In 5D, while $2\pi^3 r_s = cb_3$ and $s'/s = (1+.08^{5/3})\,\pi^3$,

in 3D, $s'/s = \pi/\varphi$ or $ds'/ds = \pi/\varphi$.

Arguably for H:

There is normally no neutron, and the 3-to-5 (requiring a 5-to-8) traverse is not present.

In that case, H can be subject to the small-value G since K is 0 (zero.)

Then H, and H alone, may become a victim of the 3D (G) gravitational energy requirement $= E_B = 680 eV/kg$ on the planet surface.

Arguably, were we constructed from 100% heavy-H (as in heavy-water where H is the single one-neutron H isotope,) then the hydrocarbons within us would not "age" because G would not apply in the presence of the stronger force K.

The three-to-five (similar in principle to the 5 to 8) traverse should represent an electron motion. These shall be beyond the scope of the present text.

MASS X VOLUME REVISITED:

As we have shown the meaning of the Bohr radius, we can now show its relation to the black-hole radius r_H per Appendices K and L.

Per the Mass / Volume (density) vs. Mass x Volume (E/E_B) diagram from Appendix E, both the atom and the black hole have small densities, but they each have opposite extremes of mass x volume (E/E_B).

We have shown the smallest possible radius r_B represents the ground state of hydrogen. This defines the smallest natural mass x volume possible in three-dimensions.

At the opposite end of the "spectrum," the largest possible mass(s) x volume(s) are represented by the m_H x r_H of black holes.

Examining the mass x volume high-and-low values at opposite ends of the "x-axis" per Appendix E, we will now use values of mass x volume having equal densities.

For the 3-dimensional intersections, we can then write:

$$(m_E + m_P) / (4/3 \; \pi \; r_B{}^3) = m_H / (4/3 \; \pi \; r_H{}^3)$$ or we can suggest (and strongly-approximate,)

$$m_P = m_H \times r_B{}^3 / r_H{}^3$$ or

$$\mathbf{m_P / m_H = r_B{}^3 / r_H{}^3.}$$

Assuming any 8-D "surface-mass" in the hole consists of a discrete value $m_N = m_P$, we can then write:

$$m_H = n_{(p+n)} m_P$$ and then,

$$\boldsymbol{n_{(p+n)} = (r_H / r_B)^3.}$$

For the figurative hyperbola at the end of Appendix E, we can then label the "top" (slope = 0) of the function Mass / Vol = f(Mass x Vol) as "NOVA."

STEREOISOMERS AND ENANTIOMERS

These are spatial variations of the same chemical compound. In particular, an enantiomer is a compound or molecular chain where the isomer is a non-superimposable mirror image of itself.

In modern medicine, it is well known that mirror-image isomers chemically behave differently from each other.

Consider the kernel groupings in the diagram below (per Appendix X.)

IMAGE				MIRROR	
	CH_3			CH_3	
CH_3		CH_3	CH_3	....	CH_3
	CH_3			CH_3	
	H_3			H_3	
	C			C	
H_3C		CH_3	H_3C	....	CH_3
	C			C	
	H_3			H_3	

While 3-dimensional chemistry should not theoretically distinguish between the mirror-image molecular groupings, the groupings are extremely different through 5-dimensional traversal.

Consider even the superimposable case of a single kernel that is bound at one molecule or chain end as CH_3 and at the opposite end as H_3C.

Consider the molecule's center of mass progressing along the curved surface in 5-dimensional space having curvature exactly per Appendix Q.

Then the CH_3 and H_3C kernels experience different spatial travel (and have very different 5-dimensional energies) similar to the longer distance of 3-dimensional space travel for a higher vs. a lower orbiting satellite.

At the start of Appendix Q, we defined the difference between "right-and-left" using the physical law and vector expression $\mathbf{F_M} = q\mathbf{v} \times \mathbf{B}$.

From over 700 million years of evolution on the planet surface, our innate systems and processes should already understand the physical and chemical meaning of right-and-left in the same way a kernel has assimilated (learned) the spatial traverse for itself.

TECHNICAL SUMMARY

1. The charge attributes + and – have been shown to be dimensional attributes of mass.

2. The semi-attribute "electron spin" has been shown to represent proton spatial position within the five-dimensional nuclear interior.

3. Quantum mechanics and the dimensional nuclear surface force F_N independently derive a universal gravitational constant.

4. The uniform gravitational constant $K(D) = K(\Delta D) + G$ where $K(\Delta D) = 7.57E+28$ (Nwt kg $^{-2}$ m ^{2}) for $\Delta D=1$.

5. $\kappa = \gamma' / \varphi'$ defines $h = h(E_B)$ and the transitional value $n = \kappa c$ as well as the transition-time expression $t = (1 + \kappa) / c$.

6. Centers of mass appear to be origins of coordinates for the sequential spatial traversal.

7. The dimensional model $t=cB$ is proved by Appendix W.

8. The five dimensional computer model has further been defined by the material-related value λ.

9. The dimensional cold fusion model has been shown. A figurative example has been suggested.

10. A kernel (CH_3) for the hydrocarbon low energy communication has been suggested.

11. Facts of dimensional neutron mass have been shown in Appendix Q and have been suggested in Appendix Y.

12. The correct nuclear radius values for He and H have been shown.

13. The value ν as in $E = h\nu$ has been redefined and re-unitized in the absence of time t.

14. We now more clearly see the meaning of $E / E_B =$ mass x volume.

15. We now more clearly see the meaning of traversal events B where one second of time t is a representation of c events B.

16. We now understand the perception of time dilation.

17. We now understand that to move forward we must first turn backward and review Diagram 21.

XIII. Reference from Prior Works

1. The Natural Theory: A Non-Mathematical Supplement

INTRODUCTION:

Our concept of continuous time is deeply rooted in our perceptions. In this segment, we consider a case where time is not in fact fully continuous and where time can be equally represented as a contiguous spatial "growth."

How Could Space Move, Progress or Sequentially "Grow"?

We start with the Fibonacci natural rate of growth. The Fibonacci spatial sequence is best Illustrated by photographs and diagrams:

A photograph of an aloe plant as it has naturally grown:

A photograph of a seashell:

A diagram of the Fibonacci spiral with curvature ρ, as it mathematically and exactly matches the natural photographs:

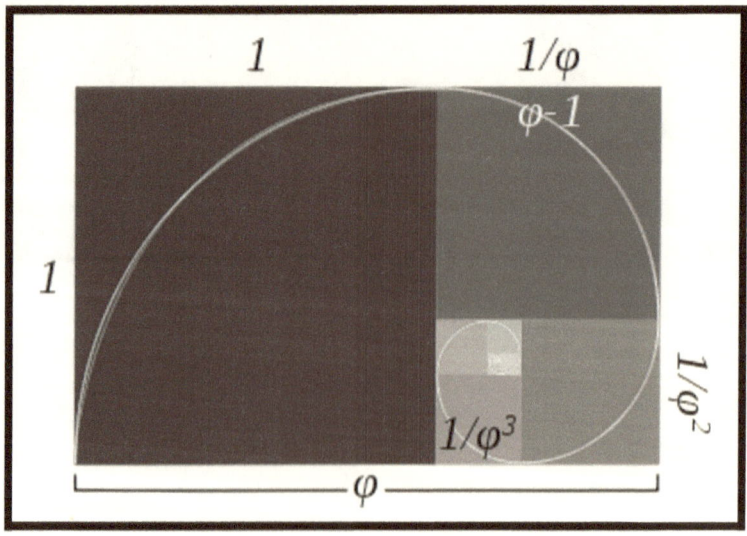

The mathematical match to nature does not stop on the planet surface:

See Appendix Supplement I for a more intuitive and mathematical treatment of the Fibonacci infinite spatial sequence.

An aside: In the year 1202, Leonardo Fibonacci was an Italian mathematician from a wealthy family. He discovered the work of an ancient (B.C.) Indian mathematician who had postulated the future population of rabbits given certain starting conditions.

We have been viewing two dimensional surfaces (photographs) from our personal three-dimensional perspective. We have seen a one dimensional line curve in two dimensions on a page of paper, on a chalkboard, or on our viewing display surfaces. We can also imagine the three dimensional result, for example the aloe plant growing nearby for us to see every day.

A CLOSER VIEW OF THE SPIRAL:

We return our attention to the diagram of the geometrical spiral having the symbol ρ in its measurements. (ρ is simply nomenclature for the ratio 0.618 defined in Appendix Supplement I.)

Since we live in three dimensions, we can easily see the two dimensional intersection within the Euclidean spiral, i.e. the linearity of the spiral intersects with three adjacent two dimensional regions per the diagram.

Space itself, as we know it, is three dimensional. If we lived in five dimensions, we could easily see the three dimensional intersection of a two-dimensional "spiral" with "five" regions of three dimensional space per the natural sequence.

We do not live in five dimensions; instead, we live in three and we cannot fully "imagine" a higher dimensional state.

An aside: If we lived in the year 1492, we may have believed the Earth was flat. We would have the same mindset as two dimensional creatures. We would have no idea that the Earth surface in fact was curved in the next highest spatial dimension (D = 3) until Columbus and others discovered they could sail ships back to where they started by never returning; instead, simply continuing on their journey brought them back to where they came from.

See Diagrams 15 and 16.

The Renaissance mariners confirmed the Earth surface was two-dimensional, but also that the surface "wrapped-around" itself in a third dimension (it is bent or curved back onto itself in three dimensional space.)

The two dimensional surface is in fact "closed" upon itself through the next higher dimension.

Time vs. Sequential Spatial Frames or Sequence of Events:

Time is definitely a great and useful concept and is a valuable measurement tool that tells us about the relative spinning position of the Earth with respect to the Sun (day vs. night) and also the relative position of the Earth's precession (slight wobble) on it's orbit around the Sun (seasons of the year.) Time defines clocks and the calendar. Time is embedded in our language(s) in terms of past, present and future tenses, expressions regarding sequences, etc. Time is embedded in our perceptions.

But time is not required to exactly match our perceptions except in the macroscopic sense. In particular, time is not required to be a "real and continuous" variable. The continuous concept of time can be replaced by a slightly contiguous view of spatial growth with the same physical result. A sequential motion of spatial frames is a new concept but yields the same macroscopic every-day result.

Anyone Could Tell the Difference?

We would never know the difference.

We have evolved on the Earth's surface over 700 million years. Everything we are and everything we think is based on our experience and evolution here. No one has the right to assume everything in the Universe needs to be the same as our Earth-bound perceptions and concepts.

To view a plot (graph) of our neurological synaptic speed, please see Appendix Supplement II. The horizontal axis units are in milli-seconds.

We do not have fast enough synapse "switching speed" to discern between sequential space (per the main text) and the concept of continuous time. Similarly, we also perceive a movie (motion picture) to be a continuous motion even though we know it is only a contiguous series of still photographs

124

placed in the right spatial order or sequence viewed quickly enough to bypass our perceptions.

Space is real. We have evolved in and through space. Mathematics comes from our surroundings, not the other way around.

While we can mathematically achieve 2 from 0 and 1, we cannot physically achieve two from nothing and something. The adjacency of 0 and 1 can physically produce only one plus one = two. Then two plus one can be subsequently achieved and so on.

Each space quantum should "experience" only each of its boundaries.

The juxtaposition of space is physical reality. The sense of time serves to approximate physical laws and works well within the bounds of our senses.

IMPLICATIONS OF QUANTUM BARRIER ENERGY:

Aging Without Time

Except as a measurement tool, there is no real physical time t; instead, time is a name for energy per sequential spatial frame.

We live (have evolved) on the planet Earth surface through our chemical composition of hydro-carbon (organic) molecules consisting of hydrogen/water and carbon/ash. We should grow and subsequently die due to the quantum energy requirement at each spatial barrier. Arguably, energy state transitions in hydrocarbon molecules are the most likely vehicles for the aging process.

For a simplified proof of hydrocarbon energy states, please see Appendix Supplement III.

The barrier energy is a quantum energy meaning the energy (including energy from us) cannot be lost unless it is an exact quantum state or multiple per the main text. An energy that is slightly different from a quantum state means $0 = $ zero energy loss. The energy (and aging) loss must be quantum (parced into finite pieces) as opposed to a continuous (could have any value) energy event.

Energy loss should effectively slow down or even stop for our physiology in the case we were located in a region of very low gravitational force. Of course, no one wants to live in a space station or on the Moon.

The Earth-surface quantum energy sum could theoretically be altered so that little or no energy was lost through the spatial progression.

ENERGY PRODUCTION

Energy production or accumulation should (arguably) be free-of-charge in the proper spatial locations of low gravitational force, for example in the region where the Moon and Earth gravity offset one another to create a very low gravitational force.

We can think about an energy transforming turbine at the Hoover Dam. With less or no E_B-energy to stop the turbine from spinning, the turbine should not need the waterfall in order to turn.

INTERACTION, ACCESS AND COMMUNICATION

With no actual physical time t, then everything progresses through the natural spatial sequence of physical events.

For example, memories may not be stored captively inside the mind; instead, the prior spatial frames (memories) are accessed through the communication ability of the brain's hydrocarbon energy states that communicate through spatial dimensions using low electro-magnetic energies. Please see Appendix Supplement IV for a technical description that is less mathematical than the main text.

DE JA VU AND ESP:

These things are not "weird"; instead, they are results of spatial intersections in the absence of time t.

For example, when you feel as though "I've been here before!" the reason you feel that way is likely because you have in fact been in that same spatial intersection "before."

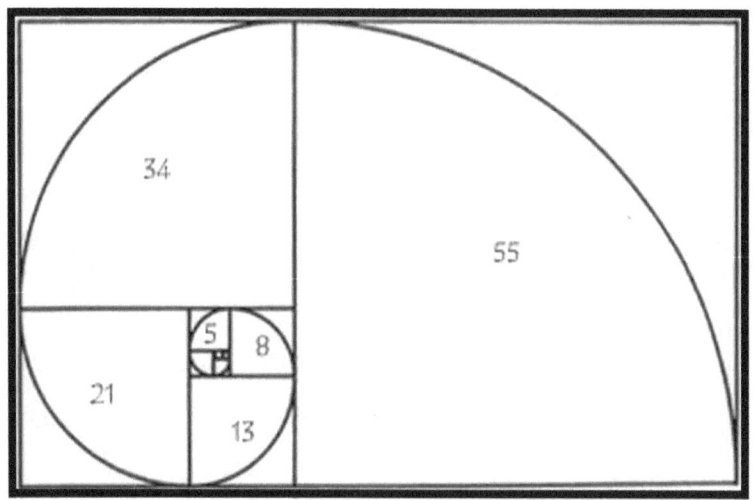

APPENDIX SUPPLEMENT I

The Fibonacci sequence is the sequence of numbers (integers) as follows:

0,1: 1, 2, 3, 5, 8, 13, 21, 34, . . . and in fact represents, among other things, a rabbit population given certain starting rules, etc. where the seed values (initial numbers) are 0 and 1 as shown, and then each subsequent number is the sum of the prior two numbers.

For example (for n = 3, meaning the third sequential number) 1 = 0 + 1.

Next, 2 = 1 + 1.

Then 3 = 2 + 1.

5 = 3 + 2, 8 = 5 + 3, and so on.

The sequence represents a start (nothing to something) followed by a growth that grows only from itself and the immediately preceding value, e.g. 3 grows from the juxta-position of itself and the only other position it is adjacent to, i.e. 2. In that case, the next sum is $3 + 2 = 5$ and so on.

The ratio, as the numbers in the sequence become large, of the given sequential number divided by the most previous sequential number is $\rho = 0.618$ . . . as approximated by 5 / 3, 8 / 5, 13 / 8, and so on. This represents the sequence of nature, i.e. the sequence of natural growth.

APPENDIX SUPPLEMENT II

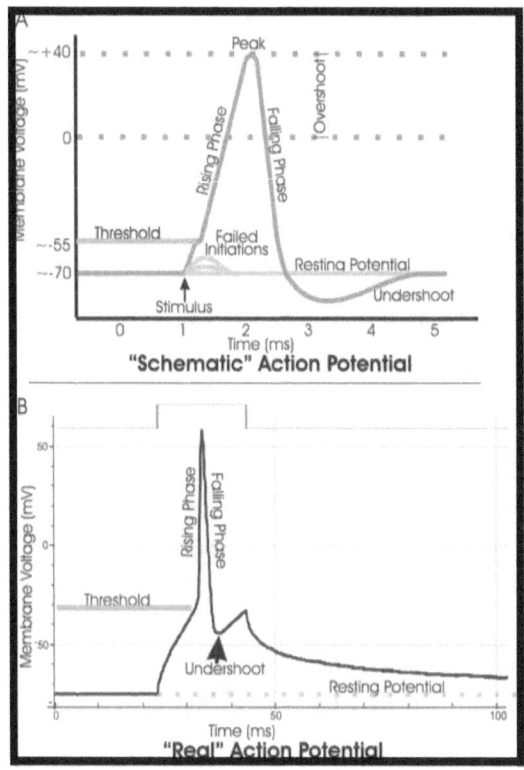

APPENDIX SUPPLEMENT III

Per the main text and Appendices D and E, the physical nature of atoms and molecules including hydrocarbon molecules and their proportional energy states (quantum mechanics) is related to mass and gravitational force within space.

Earth surface E_B (sequential boundary energy per unit mass) replaces the concept of continuous time. The energy E_B we experience is an exact quantum multiple of the known Hydrogen atomic energy states. In that case, the hydrocarbon molecular energies are susceptible to deteriorating through the quantum requirement per spatial frame.

Then life as we know it should grow and die through quantum energy states of hydrocarbon organic molecules.

Hydrocarbon death should be accelerated by a larger gravitational force (closer proximity to higher mass) while hydrocarbon extended life should be a result of less mass and a lower gravitational environment.

Common quantum energy transition sums within hydrogen alone are $13.6eV = 10.2eV + 3.4eV$ and are exact factors of the 680eV quantum energy requirement.

THEN OUR EVOLUTION IS A RESULT OF THE NATURE OF SPACE.

APPENDIX SUPPLEMENT IV

Per Appendix M in the main text:

The adjacent intersections to three-dimensional space $D = 3$ are with dimensions $D = 2$ and $D = 5$, and the intersection between three and five dimensional space is a two-dimensional surface, e.g. a closed surface around a large mass M.

The best example is the surface of the planet Earth. There is no need to consider exact altitude A from the spherical surface.

There are infinite concentric closed surfaces around the mass M, so we can define a surface "intensity" or transmission-ability proportional to r^{-2}:

Then the transmission ability T_S is proportional to $1 / r^2$ similar to the physical laws of sound, i.e. the farther away from the source you are, the less you can hear, etc.

In the absence of continuous time t, then the 3-dimensional spatial sequence progresses through 5-dimensional space. In order to access a prior 3-dimensional spatial frame, an interaction is required through a two-dimensional intersection.

As an example, the reader can think of any important memory. Subsequently, a photograph-like image immediately appears to the reader's mind. In the absence of continuous time, the three-dimensional spatial frame would need to be accessed in order to review the memory.

The access required is through a two-dimensional intersection and through five-dimensional space in order to obtain information contained in a prior three-dimensional spatial frame. The previous spatial frame(s) is still there. It didn't go anywhere. The frames should be available for continual access.

There can be access/communication at the speed of light c by energy state transitions, including hyperfine transitions, in hydrogen atoms and other molecular state transitions as well as low energy magnetic dipole moment interactions from molecular charges within large hydrocarbon (organic) molecules.

Arguably, not all memories are self-contained. Electro-magnetic communication can take place as a result of hydro-carbon molecular energy state transitions and with the surface(s) having the combined intensity $\sum T_S$ as above. The communication rate (not the perceptibility rate) should be the speed of light c.

2. Manuscript:

A MATHEMATICAL TRANSFORMATION OF VARIABLES DEFINING SPACE – TIME AND THE CONSTANT h

Marc E. King
Silicon Valley, California
June, 2012

ABSTRACT

A variable transformation for time t is supported by wave mechanics and relativity theory and shows that time and space can be related and connected by the concept of physical events per unit space. The transformation confirms our daily macroscopic experience as unchanged from classical physics while still suggests new physics regarding small energies and large spaces. Perceived time can be altered relative to

Earth-bound clocks in regions of lower or higher gravitational force. A series of calculation-verifications proves the theory, derives Planck's constant and defines quantum mechanics. Black holes and their mass-radius relationship are defined. The Schwarzschild radius is defined. Minimum and maximum energies are defined.

INTRODUCTION

In this model, it is shown that continuous time t and a contiguous view of spatial frames are mathematically the same in the macroscopic sense. A suggested transformation of variables presents interesting differences in concept for small and large energies and spaces.

CONCEPT

We postulate that continuous time as experienced can also be represented, with the same physical result, as a directional spatial sequence or frames of events.

We consider a new unit system using the transformation t = cB with c = "speed" of light, where one spatial frame (size b) is related to one physical event B by b (meter) = 1 (event) / B (events meter^{-1}.)

The transformation t = cB implies the units t (sec) = c (met/sec) x B (sec^2 met^{-1})

Then B events per meter = B sec^2 per meter, and one physical event = one square second = sec^2.

DERIVATION

From wave mechanics, we have the Schrodinger equation[i] $d\psi/dt = + -2\pi i/h$ x $E\psi$ as a partial derivative and the related approximation

$\Delta x \Delta k \geq O(1)$.

This defines the uncertainty in measurements[ii]

$\Delta x \, \Delta p \geq h/2\pi$.

Implying $\Delta p = m \, \Delta x / \Delta t$ and using a transformation for Δt, then

$\Delta p = m \, \Delta x / \Delta(cB)$

This leads to $m (\Delta x)^2 \geq (h/2\pi \text{ J-s}) (\Delta cB) = (h/2\pi \text{ J}) (cB) (\Delta cB)$

Per unit mass, then

$(\Delta x)^2 \geq (h/2\pi)(\Delta cB)(cB)$ from transformation.

For a single B (events-meter^-1) the corresponding $\Delta x = b$ meters and

$\Delta(cB) = 1/(cB)$.

Then $b \geq (\text{h-bar})^{1/2}$.

Further defining b-minimum as the minimum Δx and using the positive root in this analysis, then **b(min) =1.027E-17 meters.**

Subject to the further justification below, we assert:

$E = F\text{-sub-}B \times b$

Where $F\text{-sub-}B = F\text{-sub-}G$ = the gravitational force at the spatial location of event B.

And on the planet surface, $F\text{-sub-}B = ma = m \times 9.8$ meters-sec^{-2}.

Then $E / m = a \times b = 9.8b$ meters $/ (cB)^2$ or

$E / m = 9.8b / (c/b)^2$ and

$E / m = 9.8 / 9 (10^{-16}) b^3$ J-kg^{-1},

Or we can write the expression:

$E / m / b^3 = 1.089E\text{-}16$ J kg^{-1} or

$E\text{-sub-}B / m = 680$ eV / kg for one cubic spatial boundary.

JUSTIFICATION FOR SPATIAL DIMENSION b

Using uncertainty and similar to the derivation above, one estimation using neurological sensory communication as an upper bound on $\Delta x = b$ (in one dimension) for spatial boundary (frame) size is suggested by:

(Spatial Frame Width)$^2 \leq$ (h - bar) x (c) x (Time Required for Sensory Continuity)

Using orders of magnitude 10^{-34} J-sec (and adjusting for units) from wave mechanics and estimating the time required for sensory communication in the range 10^{-3} sec $- 10^{-6}$ sec from synapse switching (potential change) rate, we would then estimate the magnitude:

$\Delta x = b \sim 10^{-14}$ to 10^{-16} meters (for example as an upper spatial bound) in order to perceive continuity from actual contiguity of frames.

The neurological bound approximates the largest frame or spatial size that could be perceived as the continuity of time and accommodates the calculated boundary dimension $\Delta x = b \sim 10^{-17}$ meters. b(min) = 1.027E-17 meters was derived assuming t = cB so that c is assumed to be the maximum achievable velocity[iii] and as such defines a maximum sequential rate and a minimum allowable b.

JUSTIFICATION FOR SPATIAL BARRIER ENERGY

Using a one dimensional example,

E = F x distance.

We are using the transformation t = cB where B = 1 / b and b has the spatial dimension of meters.

.

The energy associated with the distance b is a function of a force F acting upon a mass m at a particular set of spatial coordinates.

It follows, the innate force acting on the mass m in space is the gravitational force.

There are no external forces to be considered for the mass m for our purpose regarding the transformation associating time and space.

VERIFICATION OF CALCULATIONS

E-sub-B is then a function of gravitational force.

Using the calculations above and for a single event B = 1, we can also write, for the planet surface as an example:

E / m = a x b = b x 9.8 (c / b)^-2 = 9.8 / c^2 / b.

Then 1.089E-16 = 9.8 / c^2 / b.

And for the planet surface E-sub-B, we verify our unit of measure calculations: b meters = (1 event / B events meter^-1) = 1.000 as a confirmation of the energy calculation 680eV / kg.

Independently, we can re-calculate the value of b using F-sub-G on the planet surface: **b = E-sub-B / m x (a)^-1 = 1.089E-16 / 9.8 = 1.111E-17 meters.**

This should be the universal value of b and is independent of F-sub-G since the accelerations "g" cancel for any spatial position.

This value is larger than the allowed minimum calculated b(min) = 1.027E-17 by the difference 8.4E-19 meters and we find the calculated surface value to be approximately 8% larger than the minimum allowed value b(min) using the Earth gravitational acceleration a = g = 9.8 m-s^-2 and using no transformations in this calculation. We do not pursue further calculations in the present scope. **(See Appendices for calculations.)**

Energy Change as a Function of F-sub-G = F-sub-B

Assuming mass m and boundary b are unchanged, then E-sub-B changes as a function of F-sub-G = F-sub-B the gravitational force at the location of physical event B.

This follows directly from

E-sub-B = F x b.

A smaller gravitational force leads to a smaller E-sub-B relative to the planet surface.

With different E-sub-B, clocks should appear to run at different rates in regions of higher or lower F-sub-G relative to the planet surface.

A fictitious force, like the Coriolis force or the weightlessness of orbit, should not affect the real force F-sub-G = F-sub-B.

CONCLUSIONS

Continuous time can be represented by a contiguous spatial sequence of frames, or boundaries $\Delta x = b$, while conforming to existing physics in the macroscopic sense and with our sensory perceptions.

The transformation t = cB leads to the spatial frame dimension b(min) = 1.027E-17 meters and corresponds to 3 x 10^8 physical events in one second of time t.

For this model,

The surface barrier energy E sub-B per unit mass = 680 eV / kg has been defined.

E-sub-B is suggested to be a function of gravitational force and so a function of spatial location.

Perceptions of Earth-time and clocks are expected to experience different rates in regions of lower or higher gravitational force relative to the planet surface.

3. A Technical Summary of *Changing Your Mind*

Per the main text and manuscript, we assert the transformation t = cB where t (sec) = c (meter/sec) x B (sec^2 / meter) = c / b (event/meter) and where a single event = one sec^2 and where we need to remove "time" t from our units of measure for non-macroscopic physics.

It has been shown that b = 1.111E-17 (meters), and that

$E_B = 680\text{eV} / 1\text{kg} / (b^3 \text{ met}^3)$.

The Fibonacci infinite sequence can be described as follows:

$\text{Lim } (n \rightarrow \text{infinity}) \ F(n-1) / F(n) = \varphi = 0.618 \ldots$ and

$\text{Lim } (n \rightarrow \text{infinity}) \ F(n-2) / F(n) = \gamma = 0.382 \ldots$.

The growth rate of space itself for three dimensions expanding through five dimensions is $r_V = \varphi^{\wedge}(5\text{-}3)$ so that

$V_5 / V_3 = e^{\wedge}(r_V)$ for one "second" of time t.

Similarly, the growth rate of the natural logarithmic base is $r_L = (1/\gamma)^{\wedge}(5/2)$ so that $e \rightarrow e^{\wedge}(r_L)$ for D=5 traversing through D=8.

Appendices A and C are attached to this text for reference.

Quantum mechanics is then described as $E / E_B = \kappa c / n$ and $h = \text{fn}(E_B) = bE_B\kappa$.

The definition of quantum mechanics becomes mass x volume and defines the chemistry of the periodic chart of elements, and the Schrödinger[2] solution is replaced by

$E' = E_B(1-p/n^2).$

Dimensional spatial curvatures are defined as $0 \leq C \leq 1$ with the terms

0 = open, and

1 = closed.

Appendix K is attached to this text as a reference for curvatures.

Appendix L is attached as references for 5-dimensional geometry and 5-dimensional nuclear force respectively.

4. Technical Summary of *Fifth Dimension*

Then there are two boundary conditions defining the natural elements:

1. E / E_B = mass x volume = atomic mass x $4/3 \, \pi \, r_A^3$
2. $b / Q_0 = r_N$ x nuclear charge / nuclear mass

Diagram 6 is the best overall summary of *Fifth Dimension*.

Appendix S is attached as reference for the added two derivative coordinates.

Appendix Q is not attached except for Diagram 18.

Other appropriate diagrams from the original Appendices Q and S are included in the Diagram Section.

APPENDIX A (REFERENCE)

The difference in value between b(min) derived from the Schrödinger equation and b calculated empirically from Earth-surface F-sub-G is 8.4E-19 meters and proves to be the mean factor 1.079 or 7.9%.

The concept of continuous time t leads to exponential growth-decay: $a = a\text{-sub-}0 \times e^{\wedge}$ (rate x time) where $e = \lim (n \rightarrow \infty) (1 + 1/n)^{\wedge}n = 2.718$.

Applying the expression for time itself, then

$$T(\text{new}) / T(\text{old}) = e^{\wedge}(r \times t) = e^{\wedge}(0) = 1.$$

For time t itself, the rate $r = 0$ and there is no change in continuous time t so that one "second" of "time t" does not change. Time t is absolute.

Differential equations for centuries, e.g. the Schrödinger equation, assume time t is a real and continuous variable.

In the transformation $t = cB$, we need to treat continuity of time as a slight-contiguity of space.

In that case, we find 3 x 10^8 met/sec to be a large enough frequency (number n) to continue using the calculated value of e = 2.718 = lim as n→infinity of $(1+1/n)^n$.

But in a directional spatial sequential model, then space itself advances or grows per some rate different from r = 0.

As a one dimensional chalk line curves in a two dimensional blackboard, and as a two-dimensional earth-surface curves in three-dimensional space, then a change in 3-dimensional space needs to take place in a mathematical dimension higher than 3.

Postulating the higher number of dimension (vertices) to be 5 as in the Fibonacci[1] infinite sequence, and considering physical events B = 1 / b = sec^2, then we calculate the following for one second of time t:

$(V\text{-sub-}S \ / \ V\text{-sub-}0)^{1/5} \ = \ e^{(r_V \ \times \ t)^{1/5}} \ = e^{(.618^{(5-3)})^{1/5}}$ = 1.079 or a 7.9% decrease in physical events B (increase in one-dimensional size b) from the continuous-time model used to calculate b(min).

APPENDIX C (REFERENCE)

A GENERAL FIBONACCI CALCULATION

The Fibonacci infinite sequence was referenced in Appendix A,

$F(n) = F(n-1) + F(n-2)$ with seed values
$F(0) = 0$ and $F(1) = 1$.

Ratios converge, and
$\lim(n \rightarrow \text{infinity})\ F(n+1) / F(n) = \varphi = (1 + 5^{1/2}) / 2 = .618\ldots$ and
$\lim(n \rightarrow \text{infinity})\ F(n-2) / F(n) = \gamma = .382\ldots$ and so on.

Writing an example expression for spatial dimension ≥ 3 per Appendix A

$\iiiint\!\!\!\iiint dV = \iiiint\!\!\!\int dV(0) \times \exp(r_V \times t)$ where
$r_V = r\text{-sub-}V = \varphi^{(D(n+1) - D(n))}$ then
$dx/dx(0) = \exp(\varphi \wedge (D(n+1) - D(n))^{1/(n+1)}$.

Except we are now doing math in another dimension, and while $e = 2.718$ in three dimensions, the base of natural logarithms should change in higher or lower dimensional space.

For example, in the case of 5 dimensions:

e → e^(1/γ)^5/2.

We quickly find dx /dx(0) = 1.08.

A different example, for the case of spatial dimension < 3:

The base e must change as a function of the power of B, i.e. in three dimensional space B ~ sec^2 while in two-dimensional space B ~ sec^3/2.

The difference in power of physical events B

2 – 3/2 = 1/2 and the two-dimensional e = 2.718^1/2.

Then we quickly find dx /dx(0) = 1.08 similar to the previous mean calculations for the difference between b(min) and b-empirical.

The (Fibonacci) calculations hold true for any spatial dimension n moving through n+1 with a dimensional adjustment for e.

APPENDIX D (REFERENCE)

PLANCK'S CONSTANT

Planck's constant[v] $h = 6.626068E-34$ met^2 kg sec^-1:

From the Schrödinger equation,

$h = 13.6eV / (1^2) / \nu = (13.6eV / (1^2) / c) \times 91.2nm$

Then $h = (E\text{-sub-}B / c) \times (91.2E-9 / 50)$

Or $h = (E\text{-sub-}B / c) \times 1.82E-9$

And $h = (E\text{-sub-}B / c) \times (b \times c / 1.82)$

So

$h = E\text{-sub-}B \times b / 1.82$ or

$h = e^{(-3/5)} \times b \times E\text{-sub-}B = b\, E_B / e^{3/5}$

where

$b = b$ for Earth surface $= 1.111E-17$ meters

and

E_B = E-sub-B = Earth surface barrier energy = 680eV/kg = 1.089E-16 J/kg

And the calculated h = $2.718^{-3/5}$ x 1.111E-17 x 1.089E-16 = 6.6E-34 per event.

More precisely from our 3-decimal calculations and per appendices A through C,

h → h(1-Δh) where Δh = $0.08^{5/2}$ and h = 6.63E-34 per event.

Units for transformed h:

h ~ met J Kg^{-1} b^{-3} ~ met^{-2} J Kg^{-1}

  ~ met^{-2} met^2 sec^{-2} Kg Kg^{-1}

  ~ $(one\ B)^{-1}$ ~ $event^{-1}$.

APPENDIX K (REFERENCE)

THE VARIOUS SIZES OF BLACK HOLES AND CURVATURES OF THREE-DIMENSIONAL SPACE

Appendix H defines the mass-radius relationship as observed in three-dimensional space: $m_H / r_H = K_G(\lambda)$ where $\lambda = \Delta\lambda = C_{R3} / C_{R5}$ and represents the ratio of curvatures from 3-dimensional space and 5-dimensional space through 8-dimensional space respectively.

We assume, for the three dimensional intersections, that $C_{R5} = 1$.

The minimum $C_{R3} = 1 / c$ and the maximum $C_{R3} = c / c = 1$.

Allowed quantum are then n / c for $n = 1$ to c.

The minimum (least dense) intersection is an intersection among 3, 5 and 8 dimensional space where $C_{R5} = 1$ and $C_{R3} = C_{R3}(min) = 1 / c$.

The next "largest" (more dense) intersection should occur for $C_{R3} = 2 / c$ and so on.

The most dense intersection occurs where $C_{R3} = c / c = 1$ and represents a closed third dimension in both eight dimensional and five dimensional space.

To visualize curvatures, the diameter of a circle = d is a straight line with curvature $C_{R1-3} = 0$ while the circumference (length πd) closes upon itself (runs into the back of itself) and has the curvature $C_{R1-3} = 1$.

The curvature C_{R2-3} is closed in 3-dimensions visualized as a spherical (or elliptical, not reviewed in this scope) surface area that has closed itself around a center-of-mass c_M.

The two dimensional surface does not alter or "grow" in three dimensions, but the one dimensional line, e.g. the straight path of a distant comet or ray of light ($C_{R1-3} = 0$) or the line of a planetary satellite $C_{R1-3} = 1$) both curve (or bend) around mass in three dimensions to the two extreme degrees of curvature.

Then the ratio $m_H / r_H = K_G$ should represent a curvature of three-dimensional space through eight-dimensional space.

APPENDIX L (REFERENCE)

DEFINITION OF MASS AND GEOMETRY FOR BLACK HOLES

From the equation $E_B = a_G$ J kg^{-1} $b_n{}^{-n}$ x b_n (= 1.089E-16 on Earth surface,) and from the definition of physical events in dimension n = B where B ~ sec^2 in 3-dimensions and sec$^{(D-1)}$ per appendix C, then b_n has the following value:

3-dimensions: b_3 = 1.111E-17 meters (per the main text)

5-dimensions: b_5 = b_3 / c = 3.703E-26 meters

8-dimensions: b_8 = b_5 / c^2 = 4.114E-43 meters and so on.

Per Appendices G, H and K, the black hole geometry is a function of mass and becomes a series of symmetrically-closed concentric surfaces having internal densities:

8-dimensional volume in 3-dimensions = 4 / 3 π r_8^3

5-dimensional volume (= $\int$ ($r_{5\text{-}8}$-to-r_5) $4\pi r$^2)

= 4 / 3 π r_5^3 (including the volume of 8)

3-dimensional volume = 4 / 3 π r_H^3 where r_H = r_5 (including volumes of 5 and 8) and the boundary condition for the most-dense black hole is then: m_H / ($4\pi r_H$^2)

= m_H / (4/3 πr_H^3) where r_H = m_H / K_G, then

m_H(max) = 3K_G kg and the corresponding r_H = 3 km.

The black hole mass m_H for the general case r_H = r_5:

$m_H(\lambda)$ = $\int$ (from r_5 − r_8—to—r_5) $4\pi r$^2 = 3$K_G(\lambda)$ where

λ = r8 / r5, r_H = r_5 and r_8 is the 3-dimensional-radius of zero-mass 8-dimensional space (b = b_8) at the center of the hole.

DERIVATIVE COORDINATES
FOR TWO ADDED VERTICES

A derivative, e.g. df(x)/dx is tangent to the curve f(x) similar to the 5-dimensional coordinates t_x and t_y from Appendix Q.

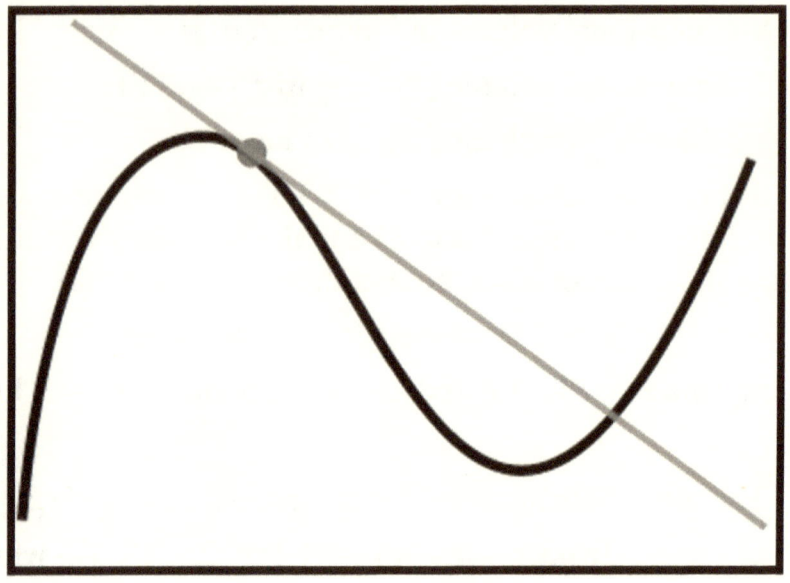

The derivative itself is independent from (orthogonal to) the point x, f(x).

The derivative is a function at the point x, f(x) relative to the points following and preceding the point x, f(x). It indicates the direction from the existing position to the next position.

In the Fibonacci spatial progression, we are traveling from the spatial dimension D=3 to D=5 and from the power-of-physical-events 2 to 3 as in the accelerations from 3 and 5 dimensional forces respectively.

The lim (n→infinity) F(n-1) / F(n) = φ = 0.618, and this is the dimensional sequential growth rate. Then, $dt_x/dx_c = dt_y/dy_c$ = φ and this matches the alternate view provided in Appendix Q.

The growth rates from Appendix C can now be more easily visualized.

To understand the rate of spatial growth using dimension 3 traversing through dimension 5 as an example: $r_v = \varphi^{(5-3)}$ = φ^2 or φ^2 meaning we have now added two extra dimensions (vertices) each having the growth rate φ so that the rate of additional volume growth is φ^2.

To understand the increasing value for 3-dimensional e = 2.718 (the base of natural logarithms) in higher dimensions D=5 and above, we need to review the dimensional geometry as follows:

Note: Here we apologize for confusion in nomenclature regarding the Fibonacci ratio and the Euler[3] angle both known as φ. Within the scope of this appendix, these will not be equated.

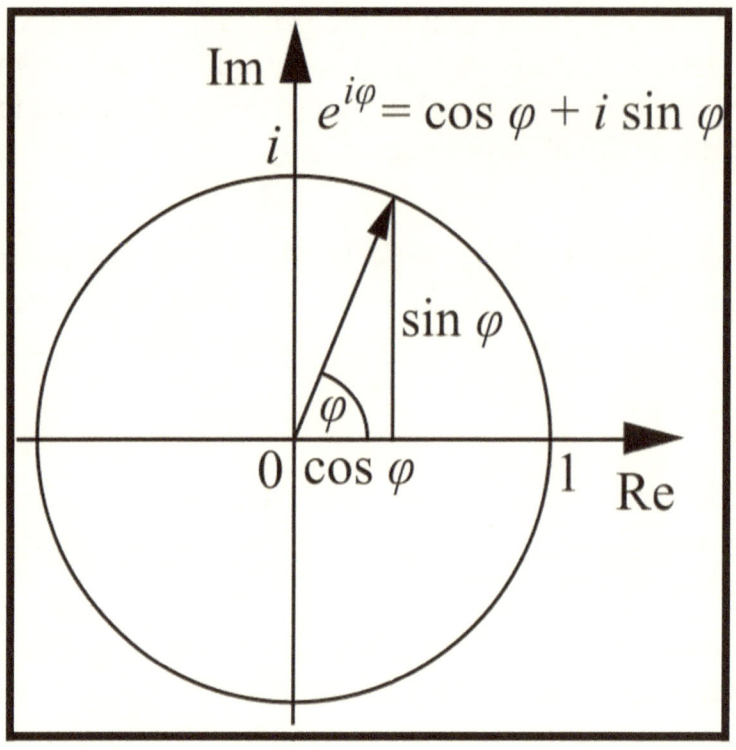

In a similar way to the rate of volume expansion and per Appendix C, the expansion of the logarithmic base (growth rate of the radius R in the above diagram) represents the addition of extra vertices. The number of vertices has grown from 3 to $3+2 = 5$. Then, $r_L = (1/\gamma)^{5/2}$ using D=5 traversing through D=8 as an example and where γ is defined in Appendix C:

$$\gamma = \lim (n \rightarrow \infty) F(n-2) / F(n).$$

The physical concept $i = (-1)^{1/2}$ or $(-1)^{1/2}$ is then represented by the example vector:

R = (0,0,0,1,0).

We can now physically extrapolate for charged points, lines, surfaces and volumes as well as their relative motions:

$\mathbf{E}_3$ is orthogonal to $\mathbf{B}_2$ where:

$\mathbf{E}_3$, $\mathbf{B}_2$, and $\mathbf{EB}_5$ represent open geometries moving within closed geometries of the same dimension, i.e. a radius within a circle and an open plane within a spherical surface respectively for $\mathbf{E}$ and $\mathbf{B}$.

We can express sequential values for $\mathbf{E}$ and $\mathbf{B}$ as below:

$\mathbf{E}$: UP 0 DN 0

$\mathbf{B}$: 0 DN 0 UP

And the corresponding vector sequence:

$\boldsymbol{EB} = \boldsymbol{EM} = (0,0,1,0,0) \ (0,0,0,1,0) \ (0,0,-1,0,0) \ (0,0,0,0,1)$ where the rate of vector progression through higher dimensional space is 1/4 cycle within b/4 meter and one complete cycle within the spatial progression b meters.

This is the nature of light.

The following diagrams represent achievable technology from the outline in appendices P through S.

These diagrams represent, among other things, the 5-dimensional computer model. The example atomic nucleus is Helium due to its multi-charge simplicity.

The smallest three dimensional energy ratio E / E_B is more than enough to provide spin state motion for the positive charges.

For example, one so-called electron-spin-split energy state transition (fine state transition $2p_{3/2}$ to $2p_{1/2}$) for Hydrogen easily provides for the real nuclear "spin" difference having the value $E = 4.5E-5$ eV with the wavelength 2.7 cm.

Energies for charge motion have already been achieved through techniques such as magnetic resonance [4].

XIV. References:

(References from prior works are not listed.)

[1] *Changing Your Mind, A Theory of Space Without Time; A Mathematical*

   Transformation of Variables. Xlibris Press, July 2012.

[2] *Younger Body Wiser Mind , The Modern Health Guide You Cannot Live*

   Without; Xlibris Press, August, 2012.

[3] **Leonardo Fibonacci**, *Liber Abbaci*; 1202.

[4] **Niels Bohr**, *On the Constitution of Atoms and Molecules*; 1913.

[5] Max Planck, *The Genesis and Present State of Development of the Quantum Theory*, 1920.

[6] William Rowan Hamilton, *Introductory Lecture on Astronomy*, 1833.